E S T A T E   P U B L

CW00787034

THANET
CANTERBURY
HERNE BAY
WHITSTABLE

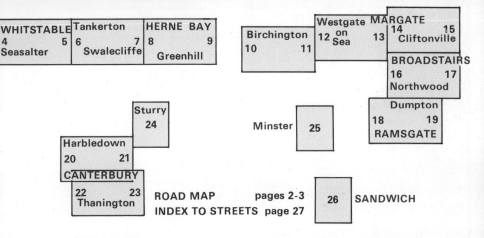

WHITSTABLE	Tankerton	HERNE BAY
4　　　　5	6　　　　7	8　　　　9
Seasalter	Swalecliffe	Greenhill

Birchington		Westgate on Sea	MARGATE	
10　　　11		12　13	14　　　15 Cliftonville	

BROADSTAIRS
16　　　17
Northwood

Sturry
24

Minster　25

Dumpton
18　　　19
RAMSGATE

Harbledown
20　　　21

CANTERBURY
22　　　23
Thanington

ROAD MAP　　　pages 2-3
INDEX TO STREETS　page 27

26　SANDWICH

One-way street　　　→
Car Park　　　　　Ⓟ
Post Office　　　　●
Public Convenience　Ⓒ
Pedestrian Precinct　▨▨▨
Scale of street plans: 4 inches to 1 mile

Street plans prepared and published by ESTATE PUBLICATIONS, Bridewell House, Tenterden, Kent and based upon the ORDNANCE SURVEY maps with the sanction of the controller of H.M. Stationery Office.

The publishers acknowledge the co-operation of Thanet and Canterbury district councils in the preparation of these maps.

ISBN 0 86084 128

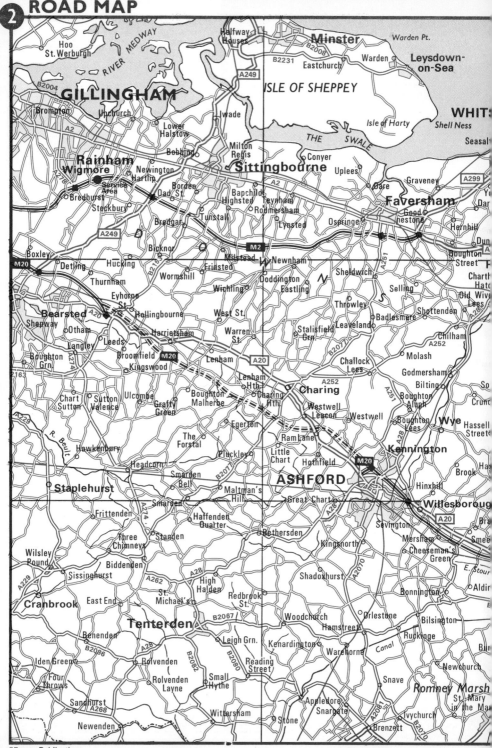

3

MARGATE
Cliftonville
Foreness Pt.
Westgate on Sea
Minnis Bay
NORTH FORELAND
B2052
B2055

HERNE BAY
Reculver
Birchington
St. Peter's
Broadstairs
Swalecliffe
ankerton
Hillborough
B2049
B2050
Acol
ISLE OF THANET
B2045
A255
A256
A254

Chestfield
Broomfield
Herne
St. Nicholas at Wade
Boyden Gate
Chislet
A299
A28
A253
Minster
A253
RAMSGATE

Hoath
Sarre
B2047
Cliffsend

Honey Hill
Broadoak
Tyler Hill
Sturry
Hersden
Grove
Westbere
Stodmarsh
Fordwich
Wickhambreux
Preston
Elmstone
Hoaden
W. Stourmouth
E.
Westmarsh
R. Stour
A291
A28
B2046
A257
A256
Great Stonar
Pegwell Bay

CANTERBURY
Littlebourne
Ickham
Wingham
Ash
Marshborough
Staple
Woodnesborough
Sandwich
Thannington
Bekesbourne
Goodnestone
Worth
A2
B2068
Patrixbourne
Bridge
Adisham
Chillenden
Knowlton
Eastry
Ham
A258

Nackington
Lower Hardres
Bishopsbourne
Kingston
Aylesham
Nonington
Bettes hanger
Northbourne
Sholden
DEAL

Bossingham
Barham
Womenswold
Tilmanstone
Elvington
Gt. Mongeham
Ripple
Walmer
B2046

tham
Derringstone
Barfreston
Woolage Green
Eythorne
E. Studdal
Sutton
Kingsdown
A260
B2065

Stelling Minnis
Denton
Shepherdswell
Coldred
Ringwould
A2

Wootton
Lydden
Whitfield
W. Langdon
E. Langdon
St. Margaret's at Cliffe
A2
A256
A258
B2058

Elmsted Court
Elham
Ewell Minnis
Temple Ewell
Guston
West Cliffe
Sth. Foreland
B2060

Rhodes Minnis
Swingfield Minnis
Densole
Buckland
W. Houghan
Alkham
DOVER

Stowting
Lyminge
Paddlesworth
Hawkinge
Capel le Ferne
A20
B2068

Postling
Etchinghill
FOLKESTONE
Stanford
Newington
A20
Saltwood
Sandgate

ymchurch
B2065
Hythe
A261
A259

ary's

PLEASE NOTE: All maps and plans contained in this publication are strictly copyright. They may not be copied or reproduced in any way without prior permission of both Estate Publications and the Ordnance Survey.

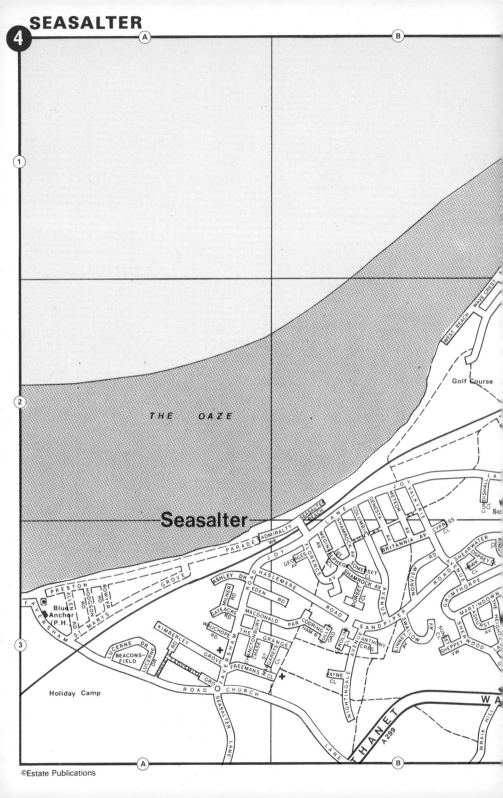

SEASALTER

4

A · B

1

2

THE OAZE

Seasalter

West Beach · Wave Crest

Golf Course

Cudishall Cl · Sc

Joy · Valkyrie · Meteor · Cypress Cl

Seasalter Beach · Lane · Columbia · Genesta Av · Britannia Av · Shearwater · Osprey Cl

Parade · Admiralty Wk · Shamrock Av · Somerset Cl · Av · Road · Avenue · Grimthorpe

Joy · Georges Av · Medina · Florence · St. Alphege Cl · Dorset Cl · Shamrock Avenue · Norrview · Martindown · Ashhurst

Grove · Ashley Dr · Haslemere · Road · Sandpiper · Swallow Av · Sunrise · Sheppey Vw

Preston · Allen Rd · Hodgson Rd · Bowyer Rd · Milner Rd · Eden Rd · Macdonald Rd · Par · Codring · Ham Rd · Grove · Carolin · Anthony Cres · Linnet Av · Sherwood · Sheppey Vw

Faversham · St. Marys · Gateacre Rd · Wauchope Rd · The Grange · St. Margarets Cl · Avenue · Wraik Hill

Blue Anchor (P.H.)

Kimberley Rd · Chanctonbury Chase · Freemans Cl · Nightingale Cl · Jayne Cl

Lucerne Dr · Beacons-field · Lucerne Dr · Grove · Faversham Gro

Holiday Camp

Road · Church · Seasalter Lane · Thanet · Wa · A 299

B

A

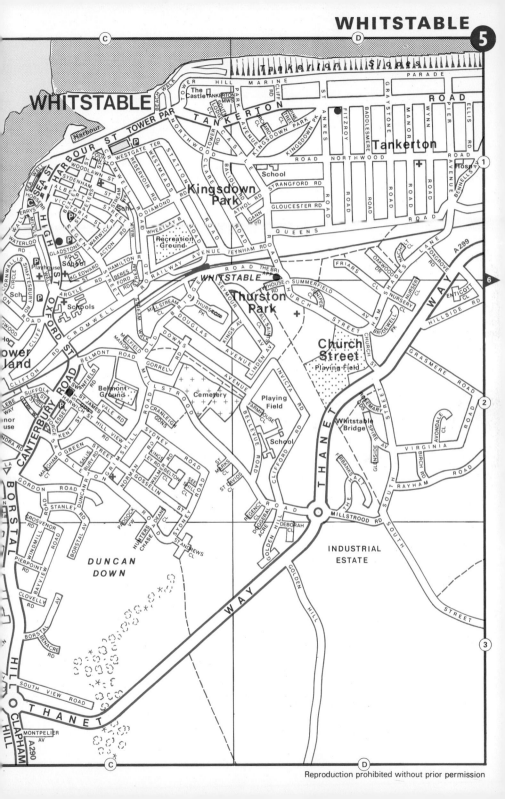

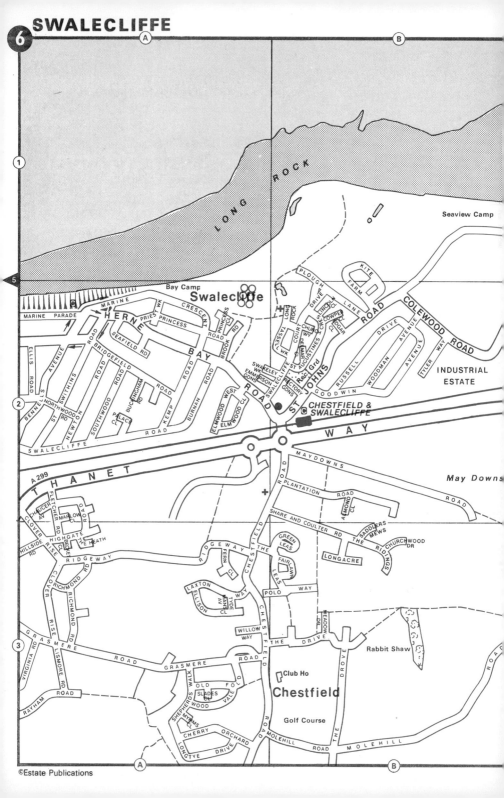

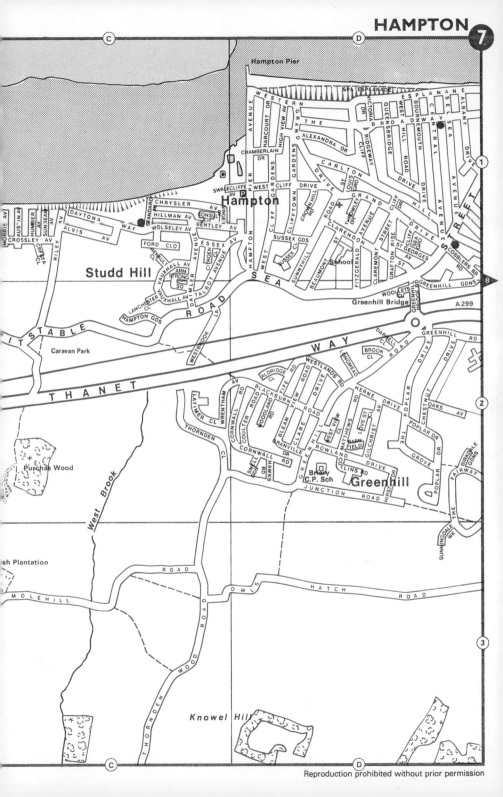

Hampton Pier

SEA ESPLANADE

Hampton

Studd Hill

THANET

Caravan Park

Purchas Wood

West Brook

sh Plantation

Greenhill

Greenhill Bridge

A299

Knowel Hill

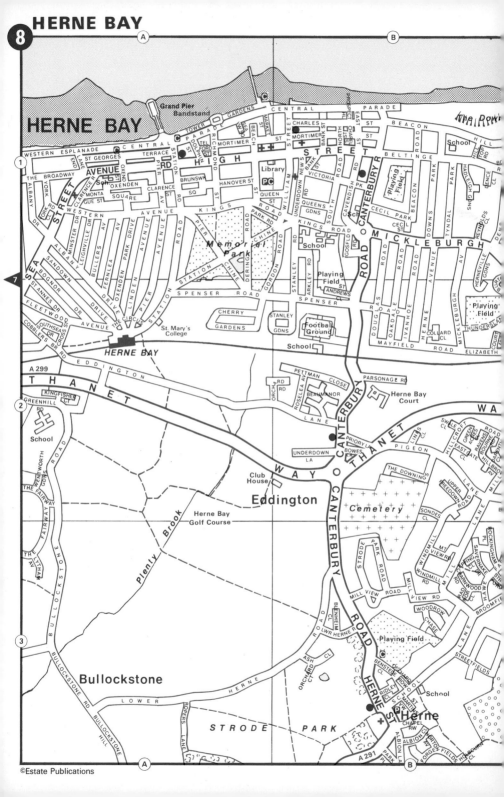

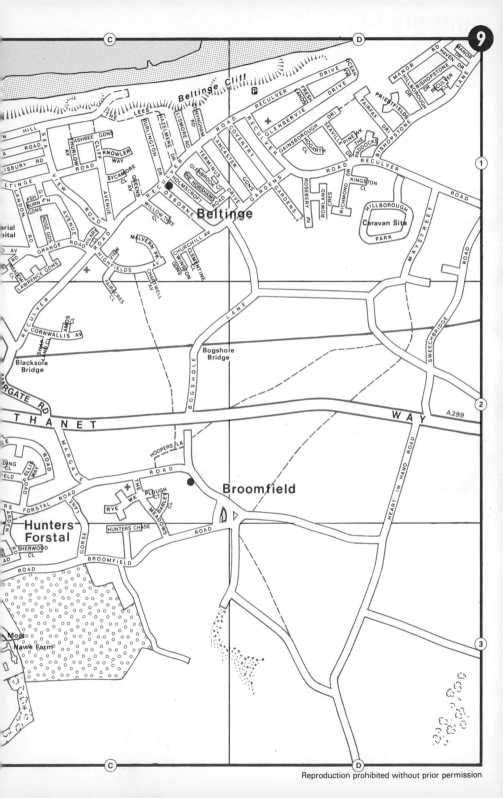

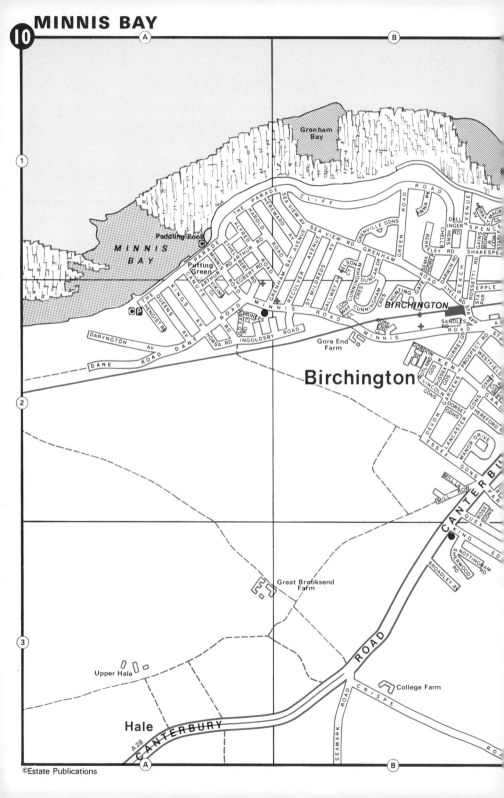

MINNIS BAY

Grenham Bay

MINNIS BAY

Paddling Pool

Putting Green

BIRCHINGTON

Gore End Farm

Birchington

Great Brooksend Farm

Upper Hale

College Farm

Hale

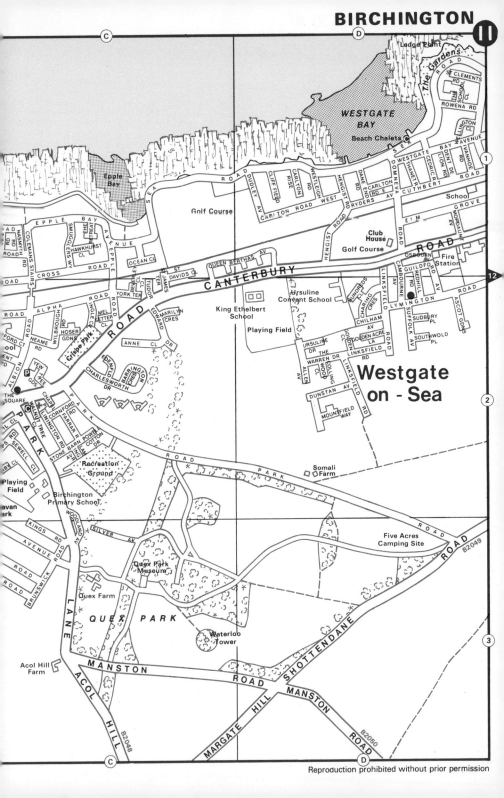

Westgate
on - Sea

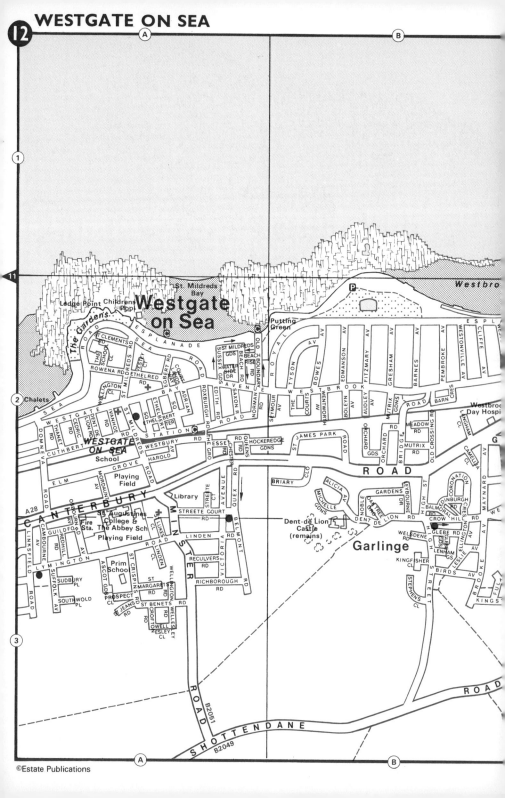

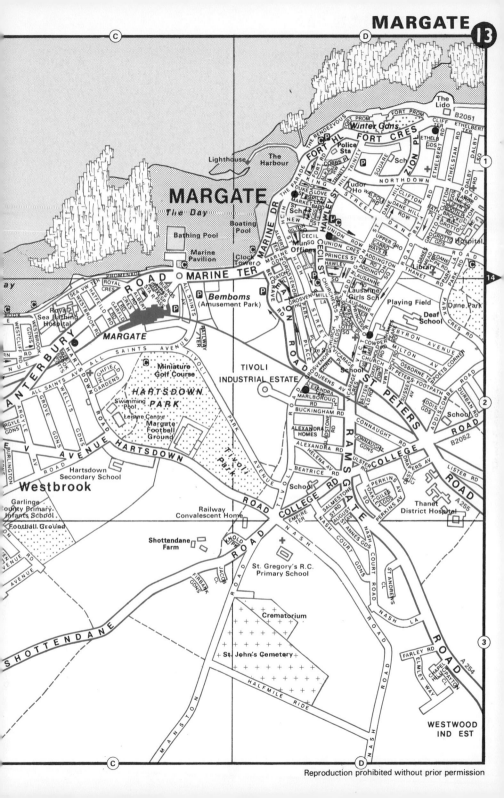

MARGATE

13

CLIFTONVILLE

©Estate Publications

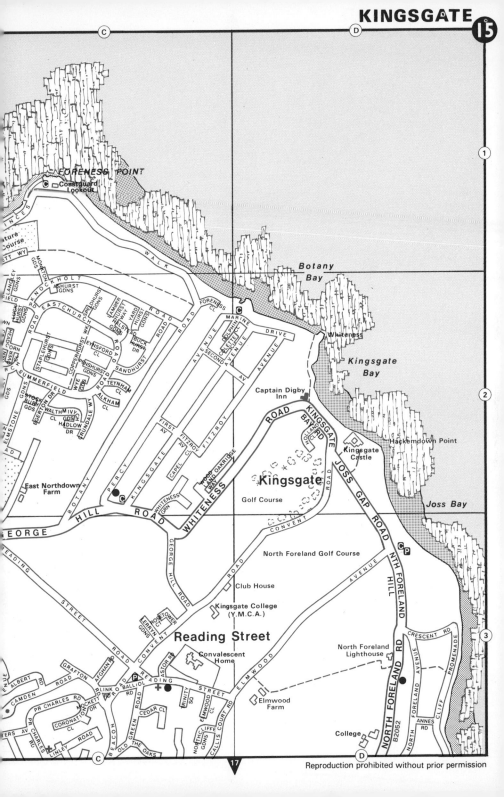

FORENESS POINT

Coastguard Lookout

Botany Bay

Whiteness

Kingsgate Bay

Captain Digby Inn

Kingsgate Castle

Hackemdown Point

Kingsgate

Golf Course

Joss Bay

East Northdown Farm

GEORGE HILL ROAD

WHITENESS ROAD

Convent

North Foreland Golf Course

Club House

Kingsgate College (Y.M.C.A.)

NTH FORELAND HILL

Reading Street

North Foreland Lighthouse

Crescent Rd

Promenade

Convalescent Home

Elmwood Farm

College

NORTH FORELAND RD

B2052

North Foreland Avenue

Annes Rd

Cliff

NORTHWOOD

16

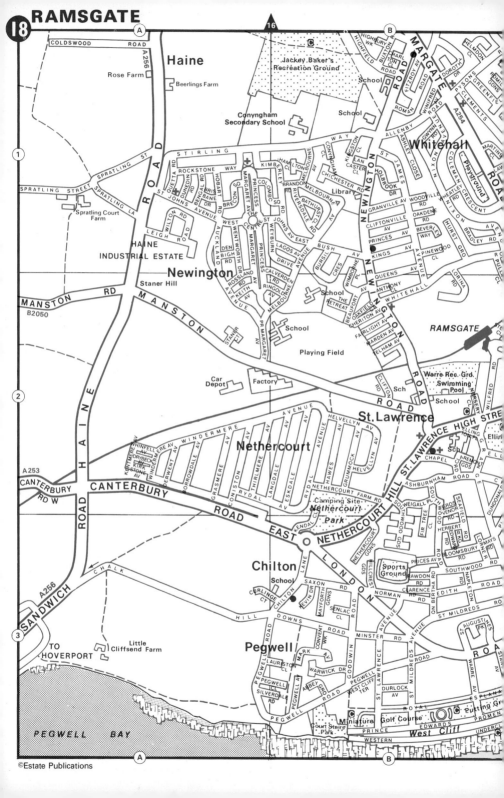

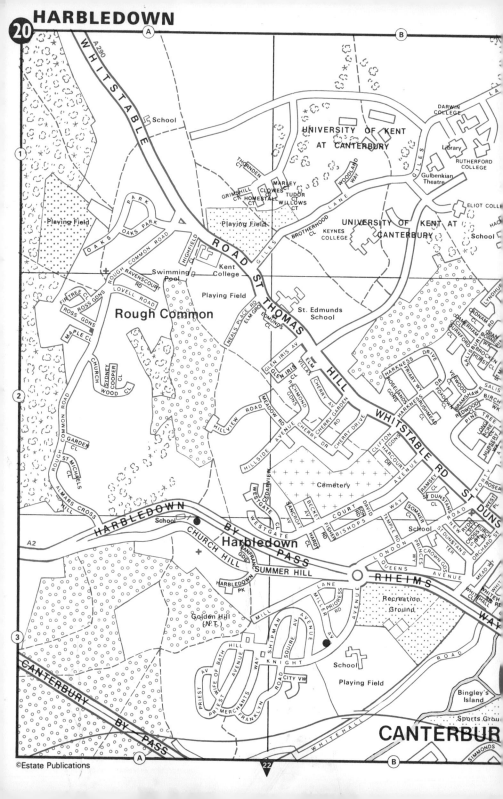

A290

WHITSTABLE

School

A

B

UNIVERSITY OF KENT AT CANTERBURY

DARWIN COLLEGE

Library

RUTHERFORD COLLEGE

Gulbenkian Theatre

ELIOT COLL

THORNDEN

MARLEY

GRIMSHILL CT

CLOWES CT

HOMESTALL CT

TUDOR WILLOWS CT

WOODLAND WAY

GILES LANE

School

1

Playing Field

PARK

OAKS PARK

OAKS

Playing Field

Playing Field

BROTHERHOOD CL

KEYNES COLLEGE

UNIVERSITY OF KENT AT CANTERBURY

HIGHFIELD

ROUGH COMMON ROAD

RAVENSCOURT RD

Swimming Pool

Kent College

ROAD ST THOMAS HILL

St. Edmunds School

FIR TREE CL

ROSS GDNS

LOVELL ROAD

ROSS GDNS

Playing Field

NEALS PLACE RD

ELM GRO

EDMUNDS CL

HARKNESS DRIVE

ROSELANDS GDNS

FRIARY WY

TRISHMEAD

PATERHAM BOURNE

GURNFORD

BROCKEN

BIRCH

SALIS

MAPLE CL

Rough Common

CHURCH WOOD

SIDNEY COOPER CL

HILLVIEW ROAD

MEADOW RD

RICHMOND GDNS

ELM VILLAS

GLEN IRIS AV

GLEN IRIS

CHERRY AV

CHERRY DR

CHERRY GARDEN RD

CHERRY DRIVE

VELWOOD

PINE TREE

REDWOOD

NURSERY

2

ROUGH COMMON ROAD

GARDEN CL

ST MICHAELS CL

MILLSIDE AVENUE

CLIFTON GDNS

HARBLEDOWN AVENUE

HARCOURT DR

ALMARS CROSS HILL

HARBLEDOWN

School

CEDARVIEW

WESTGATE

BANCROFT

BECKET

SANCROFT HARD

FISHER

CLOSE

COURT

BISHOPS

TEMPLE RD

DAVID

ROAD

SOMNER WAY

Cemetery

ST DUNSTANS CL

RAMSEY WAY

WHITSTABLE RD

ST DUNS

School

A2

CHURCH HILL

Harbledown

HARBLEDOWN PK

SUMMER HILL

BY PASS

LANGTON

GODS

LONDON

DUNSTAN'S

PRINCES

CROWN GDNS

NEW RUSE

ORCHARD RD

RHEIMS

QUEENS

AVENUE

Recreation Ground

WHIT

WHICH

HILL

WAY

3

CANTERBURY BY-PASS

Golden Hill (N.T.)

PRIEST AV

WIFE OF BATH HILL

MERCHANTS

AVENUE

FRANKLIN

KNIGHT WAY

SHIPMAN WAY

SQUIRE AV

CITY VW

MILLER AV

MILL LANE

PRIORESS RD

AVENUE

School

Playing Field

Bingley's Island

Sports Grou

WAY

CANTERBUR

WHITEHALL

SIMMONDS

A

B

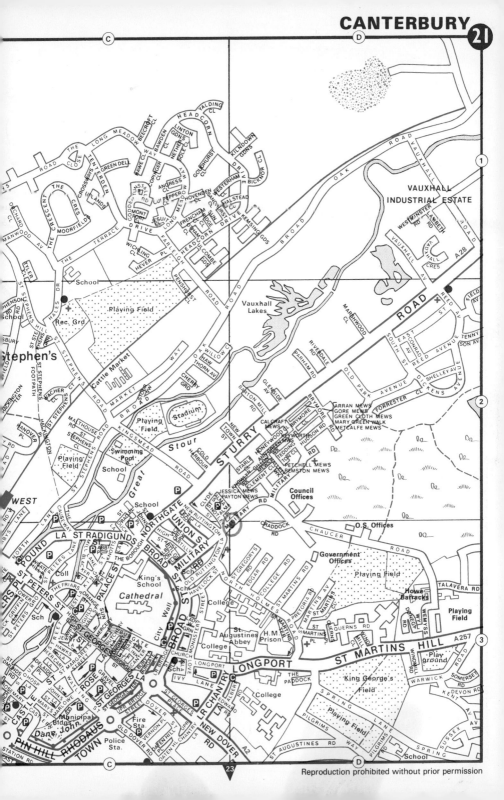

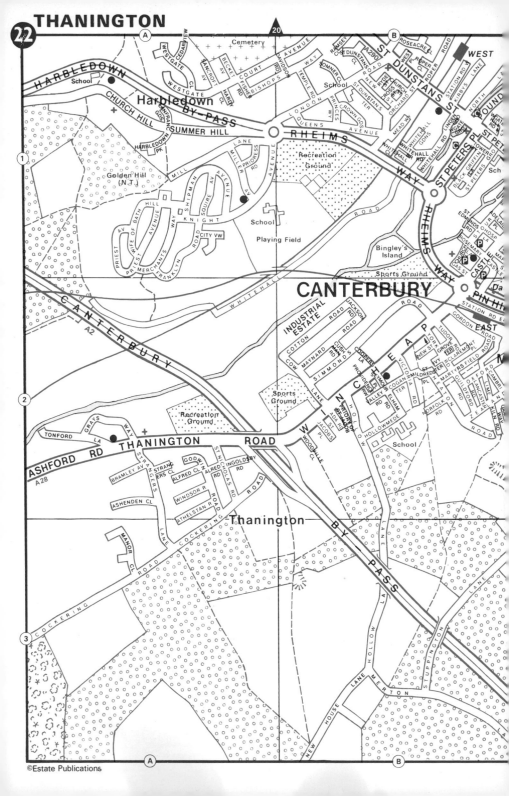

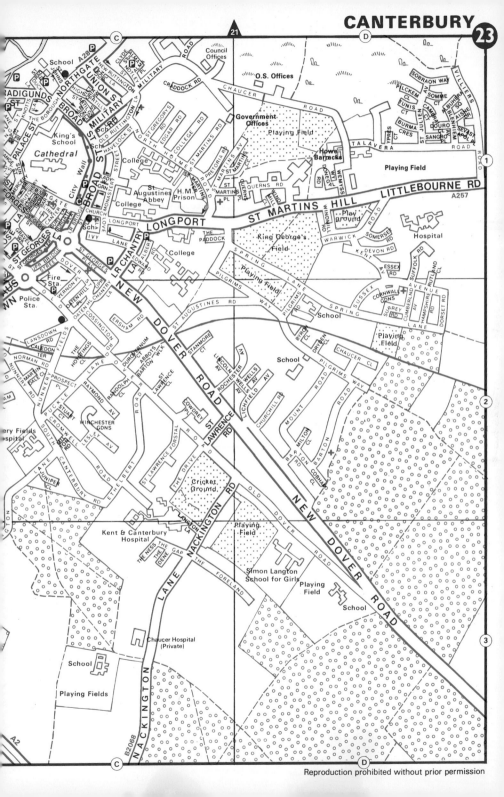

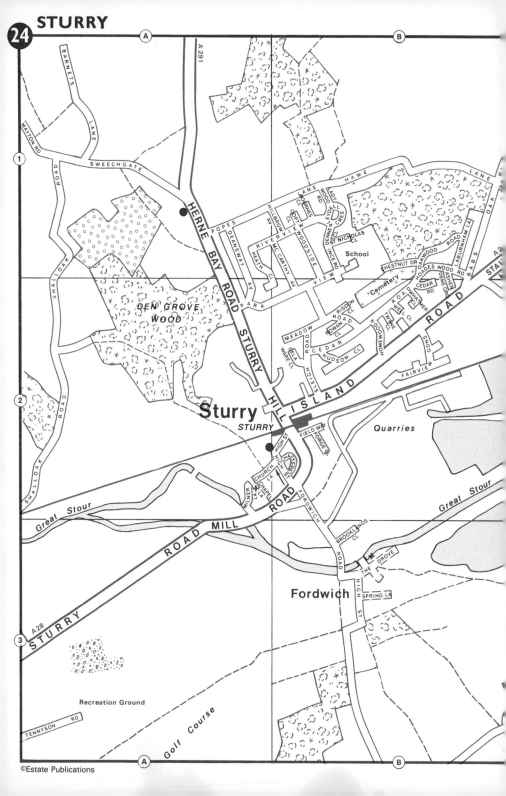

A   B

1

2

3

Sturry
STURRY

Fordwich

DEN GROVE WOOD

HERNE BAY ROAD

STURRY HILL

ISLAND ROAD

STURRY ROAD

MILL ROAD

A28

A291

BARNETS LANE

MAYTON RD

SWEECHGATE

SHALLOAK ROAD

School

Cemetery

Quarries

Great Stour

Recreation Ground

Golf Course

TENNYSON RD

ST NICHOLAS CL

HAWE LANE

OAK

CHESTNUT DR

OAKWOOD RD

LABURNHAM LA

BABS

FAIRVIEW

THE DROVE

SPRING LA

HIGH ST

BROOKLANDS CL

FORDWICH ROAD

HUDSON CL

CEDAR RD

MEADOW ROAD

RIGDON ROAD

ROWAN CL

HOMEWOOD

OAKHAMMER CL

POPES

DEANSWAY

HEATH CL

RIVER VIEW

MC CARTHY AV

PARK VIEW

HILLBROW AV

WOODSIDE

ST NICHOLAS RD

DENNE CL

EYDELL CRES

LADY WOOTTONES GREEN

LANE

HOADES WOOD RD

CEDAR

WARE CL

DELA

SLEIGH ROAD

HIGH ST

CHURCH LA

CHAPEL LA

MILNER LA

CHA CRES

FIELD WAY

FORGE

STATION RD

©Estate Publications

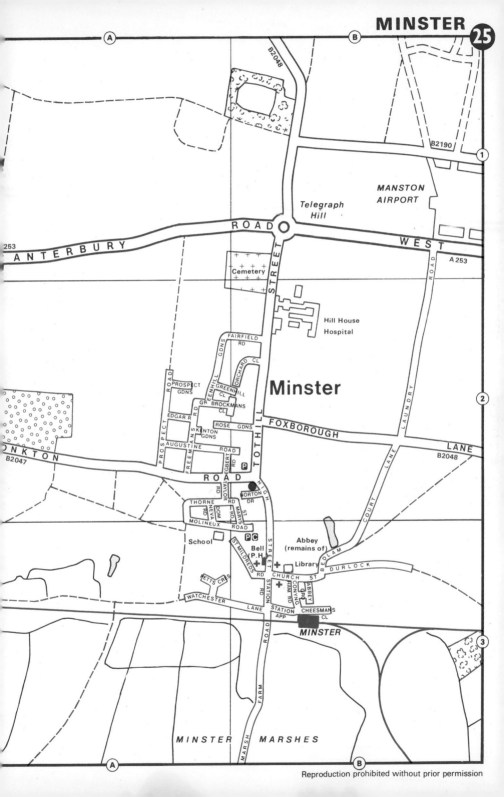

B2048

A

B

B2190

1

MANSTON
AIRPORT

Telegraph
Hill

253

ROAD   WEST

ANTERBURY

A 253

Cemetery

STREET

Hill House
Hospital

FAIRFIELD
RD

Minster

ORCHARD CL

GREENHILL CL

PROSPECT
GDNS

GDNS

ROAD

GREENHILL

BROCKMANS
CL

ROSE GDNS

FOXBOROUGH

LANE

EDGAR R

KENTON
GDNS

TOTHILL

LANE

AUGUSTINE

FREEMANS

ROAD

B2048

ONKTON

PROSPECT

LAURY

B2047

EGBERT RD

HIGH

COURT

P

NORTON
DR

ST

ROAD

THORNE RD

DOWNE

VANE RD

MARYS RD

MOLINEUX

ROAD

STREET

Abbey
(remains of)

BEDLAM

DURLOCK

School

P C

Bell
P.H.

ST MILDREDS RD

Library

CONYNG

HETTS CRES

CHURCH   ST

FARM RD

ABBEY GRO

WATCHESTER   LANE

STATION

CONYNG

ROAD

STATION
APP

CHEESMANS
CL

MINSTER

3

FARM

MARSH

MINSTER   MARSH   MARSHES

A

B

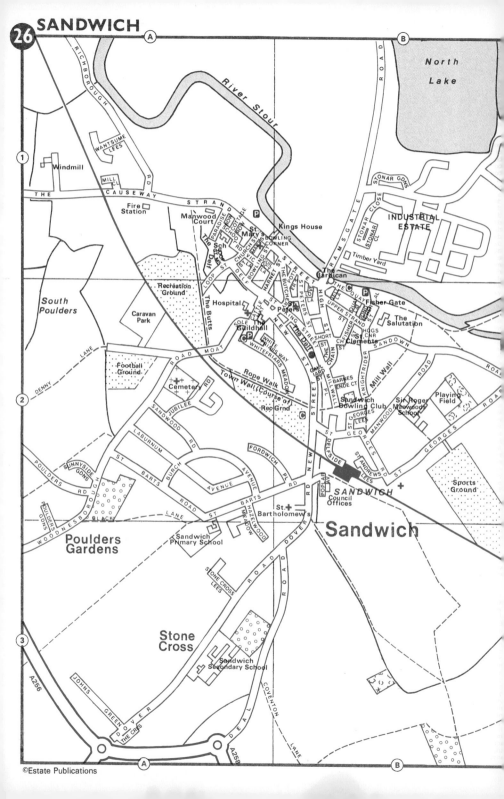

North Lake

RICHBOROUGH ROAD

River Stour

STONAR ROAD

WANTSUME LEES

MILL CL
MILL RD

Windmill

THE CAUSEWAY

STRAND

Fire Station

PARADISE ROW
SCHOOL RD
CHURCH RD
ST. MARY'S RD
DELF
LOOP
The Dell
The Butts

Manwood Court

P

St. Mary's
Sch

Kings House

Bowling Corner

STONAR GDNS
STONAR CLOSE
STONAR CL

INDUSTRIAL ESTATE

Timber Yard

The Barbican

The Quay

Fisher Gate

Fisher Gate

The Salutation

HOGS CNR
St. Clements

STONAR ROAD

South Poulders

Recreation Ground

Caravan Park

Hospital

Guildhall

St. Peters

NEW STREET
HARNET ST
THE BUTCHERY
ST. PETERS ST
UPPER STRAND ST
HIGH ST
SHORT ST
CHAIN ST
QUAY LANE
GALLARD ST

KNIGHTRIDER ST

SANDOWN

The DELF

WHITE FRIARS WAY
WHITEFRIARS MEAD
Rope Walk
Town Wall (Course of)
Rec Grnd

MOAT ROAD
JUBILEE RD
SANDWOOD RD

Football Ground

Cemetery

C

BARNES ENDE CT
Sandwich Bowling Club
GEORGES LEES

Sir Roger Manwoods School

Mill Wall

MILL WALL

NEW STREET

DELFSIDE

POPLAR RD

NEW STREET RD

DOVER ROAD

Playing Field

GEORGES RD

MANWOOD RD

ST. ANDREWS RD
ST. GEORGES RD

Sports Ground

SUNNYSIDE GDNS
POULDERS GDNS RD
POULDERS RD
WOODNESBOROUGH ROAD

LABURNUM ST
BARTS BURCH ST
BARTS LANE
ST. BARTS RD
AVENUE
FORDWICH PL
HAZELWOOD MEADOW

St. Bartholomew's

Council Offices

SANDWICH

Sandwich

Poulders Gardens

BLACK

Sandwich Primary School

Stone Cross

STONE CROSS LEES

Sandwich Secondary School

DENNY LANE

A256

JOHNS
GREEN
THE CRES
DOVER ROAD

A258

DEAL ROAD

COVENTON LANE

©Estate Publications

INDEX TO STREETS

...ds marked * do not
...ear on the map, due to
...k of space.
...ir location will be found
...he vicinity of the name
...owing them in the
...ex.

bey Gro, Minster 25 B3
bey Gro, Pegwell 18 B3
bots Barton Walk 23 C2
bots Hill 19 C3
bots Pl 21 C3
ol Hill 11 C3
on Rd 5 C1
d Rd 22 B2
dington Pl 19 C3
dington Rd 13 D1
dington Sq 13 D1
dington St, Margate 13 D2
dington St, Ramsgate 19 C3
discombe Gdns 13 D2
discombe Rd 14 A3
elaide Gdns 19 C3
elaide Pl 21 C3
miralty Walk 4 A3
rian Sq 12 A2
han Rd 17 C1
edale Rd 14 A2
ne Dri 23 D1
any Rd 7 D1
ert Rd, Canterbury 23 C1
ert Rd, Margate 13 C2
ert Rd, Ramsgate 19 D2
ert Rd, St Peters 17 C1
ert St, Ramsgate 19 C3
ert St, Whitstable 5 C1
ert Ter 13 D1
ion Clo 8 B3
ion Hill 19 C3
ion La 8 B3
bion Mews,
Cottage Rd 19 C3
ion Pl, Canterbury 21 C3
ion Pl, Ramsgate 19 C3
ion Rd, Birchington 11 C2
ion Rd, Cliftonville 14 A2
ion Rd, Ramsgate 19 D2
ion Rd, St Peters 17 C2
uhera Sw 23 D1
erney Gdns 17 C2
dridge Clo 7 D2
exandra Dri 7D1
exandra Homes 13 D2
exandra Rd,
Birchington 10 B3
exandra Rd,
Broadstairs 17 D2
exandra Rd, Margate 13 D2
exandra Rd,
Ramsgate 19 C2
exandra Rd, Whitstable 5 C2
exandra Ter 13 D2
red Clo 22 A2
red Rd, Birchington 10 A1
red Rd, Cliftonville 14 A2
red Rd, Thanington 22 A2
cia Av 12 B2
kali Row 13 D1
kham Clo 15 C2
Saints Av 13 C2
Saints Clo 5 D2
en Av 11 D2
en Rd 4 A3
enby Rd 18 B1
lliance Rd,
Elizabeth 19 D3
na Pl, Canterbury 21 C2
na Pl, Ramsgate 19 C2
na Rd, Beltinge 9 C1
na Rd, Margate 13 D2
na Rd, Ramsgate 19 C2
na St 21 C2
mond Clo, Northwood 16 B3
mond Clo, Swalecliffe 8 B2
pha Rd, Birchington 10 B2
pha Rd, Ramsgate 19 C3
vis Av 7 C1
mherst Clo 14 B3
nos Clo 9 C2
gress Clo 21 C1
n Green Walk 21 D2
na Pk 10 B1
ne Clo 11 C2
nes Rd 15 D3
ns Rd 19 C2

Anson Clo 16 B3
Anthony Clo 18 B2
Anthony Cres 4 B3
Anvil Clo 11 C2
Appledore Clo 14 B3
Approach Rd,
Broadstairs 17 C3
Approach Rd,
Cliftonville 14 A2
Archway 19 C3
Arden Rd 9 C2
Argyle Av 13 C2
Argyle Gdns 13 C2
Argyle Rd 5 C2
Arkley Rd 8 B2
Arklow Sq 19 D2
Arlington Gdns 14 B3
Armstrong Sq 7 C1
Arnold Rd 13 D2
Arran Mews 21 D2
Arthur Rd, Birchington 10 A2
Arthur Rd, Cliftonville 14 A2
Artillery Gdns 21 C3
Artillery Rd 19 C2
Arundel Rd 14 A2
Ascot Gdns 12 A3
Ash Clo, Broadstairs 16 B2
Ash Clo, Herne 8 B3
Ashbee Gdns 9 C1
Ashburn Gdns 9 C1
Ashburnham Rd 18 B2
Ashdown Clo 8 B3
Ashenden Clo 22 A2
Ashendene Gro 24 B1
Ashford Rd 22 A2
Ashley Clo 18 B1
Ashley Dri 4 A3
Ashtrees 8 B2
Ashurst Av 4 B3
Ashurst Gdns 15 C2
Astor Rd 17 C1
Athelstan Rd, Margate 13 D1
Athelstan Rd,
Thanington 22 A2
Athol Rd 5 D1
Auckland Av 18 A1
Audley Av 12 B2
Augusta Rd 19 D2
Augustine Rd 25 A2
Austin Av 7 C1
Avebury Av 19 D2
Avenue Gdns 14 B2
Avenue Rd, Herne Bay 8 A1
Avenue Rd, Ramsgate 19 D2
Avondale Clo 5 D2
*Avondale Pl, Market St 13 D1
Ayton Rd 19 C3

Babs Oak Hill 24 B1
Baddlesmere Rd 5 D1
Bairds Hill 17 C2
Bairdsley Clo 17 C2
Balcomb Cres 14 B3
Baliol Rd,
Reading Street 17 C1
Baliol Rd, Whitstable 5 C1
Balmoral Pl 19 D2
Balmoral Rd 12 B2
Bank St 8 B1
Barley Clo 9 C2
Barn Cres 12 B2
Barnes Av 12 B2
Barnesende Ct 26 B2
Barnets La 24 A1
Barnfield 7 D2
Barrington Cres 11 C2
Barrows Clo 10 B2
Barton Mill Rd 21 D2
Barton Rd 23 D2
Bath Pl 13 D1
Bath Rd 13 D1
Bathurst Clo 18 B1
Bawden Clo 21 C1
Bay View Rd 19 D1
Bayview Rd 5 C3
Beach Av 10 B2
Beach Rise 12 A2
Beach Rd 12 A2
Beach St 8 A1
Beach Walk 5 C1
Beacon Av 9 C1
Beacon Hill 8 B1
Beacon Rd, Herne Bay 8 B1
Beacon Rd, St Peters 17 C2
Beaconfields 4 A3
Beaconsfield Gdns 17 C2
Beaconsfield Rd 21 C2

Beatrice Rd 13 D2
Beaufort Av 18 B2
Beaumanor 8 B2
Beaumont St 7 D2
Becket Av 20 B2
Becket Clo 6 A3
Bedford Ct 17 D2
Bedlam Court La 25 B3
Beech Dri 16 B3
Beechcroft Gdns 19 D2
Beecroft Clo 21 C1
Beer Cart La 21 C3
Belgrave Clo 19 C2
Belgrave Rd 13 D2
Bell La 26 B2
Bellevue Av 19 C2
Bellevue Rd, Ramsgate 19 D2
Bellevue Rd, Whitstable 5 D2
Bellevue Rd, Broadstairs 17 D2
Belmont Rd, Ramsgate 19 C2
Belmont Rd, Westgate 12 A3
Belmont Rd, Whitstable 5 C2
Belmont St 19 C2
Beltinge Rd 9 B1
Belton Clo 5 C2
Belvedere Rd 17 D3
Benacre Rd 5 C3
Bengal Rd 18 A1
Bennells Av 6 A2
Benstede Clo 8 B3
Bentley Av 7 C1
Beresford Gdns 14 B1
Beresford Rd, Ramsgate 19 C3
Beresford Rd, Whitstable 5 C1
Berkeley Rd 10 B1
Best La 21 C3
Bethesda St 19 C2
Beverley Clo 11 C2
Beverley Rd 21 C2
Beverley Way 18 B1
Bexley St 5 C1
Bicknor Clo 21 D1
Biddenden Clo 14 B2
Bilton Sq 13 D1
Birch Clo 16 B3
Birch Rd 5 D2
Birchwood Walk 20 B2
Birds Av 12 B3
Birkdale Gdns 7 D2
Bishops Av 17 D1
Bishops Way 20 B3
Bishopstone Dri 9 D1
Bishopstone La 9 D1
Black Griffin La 21 C3
Black La 26 A2
Blackburn Rd 7 D2
Blackfriars St 21 C3
Blean View Rd 7 D2
Blenheim Av 23 D1
Blenheim Clo, Herne 8 B3
Blenheim Clo,
Northwood 16 B3
Bloomsbury Rd 18 B3
Bognor Dri 8 A1
Bogshole La 9 C2
Boleyn Av 12 B2
Bolton St 19 C2
Booth Pl 13 D1
Borrowdale Av 18 A2
Borstal Av 5 C3
Borstal Hill 5 C2
Botany Rd 15 C2
Boughton Av 19 D1
Boundary Rd 19 C2
Bournemouth Dri 7 D1
Bournes Clo 24 B1
Bowes Av 12 B2
Bowes La 8 B2
Bowland Clo 9 C3
Bowling St 26 A1
Bowyer Rd 4 A3
Bracken Ct 17 C2
Bradley Rd 18 B1
Bradstow Way 17 C2
Bramblefield Clo 8 B2
Bramley Av 22 A2
Bramshaw Rd 20 B2
Brassey Av 17 C3
Brecon Sq 18 A1
Briars Walk 17 C3
Briary Clo 12 B2
Bridewell Park 5 D2
Bridge Rd 12 B2
Bridgefield Rd 6 A2
Brights Pl 19 C2
Brisbane Dri 18 A1
Bristol Rd 23 C2

Britannia Av 4 B3
Broad Oak Rd 21 C2
Broad St, Canterbury 21 C3
Broad St, Margate 13 D1
Broad St, Ramsgate 19 C2
Broadley Av 10 B3
Broadley Rd 16 B1
Broadstairs Rd 16 B2
Brockenhurst Clo 20 B2
Brockenhurst Rd 19 D2
Brockley Rd 13 D1
Brockmans Clo 25 A2
Bromstone Rd 17 C3
Brook Clo 7 D2
Brook Rd 6 A2
Brooke Av 12 B3
Brooklands Clo 24 B3
Broomfield Rd 8 B3
Brotherhood Clo 20 B1
Brunswick Rd 11 C3
Brunswick Sq 8 A1
Brunswick St 19 C2
Brymore Clo 21 D2
Brymore Rd 21 D2
Buckhurst Dri 15 C2
Buckingham Rd,
Broadstairs 17 D3
Buckingham Rd,
Margate 13 D2
Buckingham Rd,
Swalecliffe 6 A2
Buenos Ayres 13 C2
Buller Gro 4 B3
Bullers Av 8 A1
Bullockstone Hill 8 B3
Bullockstone Rd 8 A3
Burch Av 26 A2
Burgate 21 C3
Burgate La 21 C3
Burlington Dri 9 C1
Burlington Gdns 13 C2
Burma Cres 23 D1
Burnan Rd 6 A2
Bursil Cres 18 B1
Burton Fields 8 B2
Bush Av 18 B1
Butchery La 21 C3
Buxton Rd 16 A3
Byron Av 13 D2
Byron Clo 23 D2

Caernarvon Gdns 17 D2
Calcraft Mews 21 D2
Caledon Ter 23 C2
Callis Court Rd 17 C2
Calverden Rd 18 B2
Cambourne Av 12 A3
Cambrai Ct 23 D1
Cambridge Rd 22 B2
Cambridge Way 22 B2
Camden Rd, Ramsgate 19 C3
Camden Rd, St Peters 17 C1
Camden Sq 19 C2
Camellia Clo 12 B2
Cannon Rd 19 C2
Cannonbury Rd 19 C3
Canterbury Clo 16 B2
Canterbury La 21 C3
Canterbury Rd East 18 A2
Canterbury Rd West,
Chilton 18 A2
Canterbury Rd West,
Minster 25 A1
Canterbury Rd,
Birchington 10 A3
Canterbury Rd,
Herne Bay 8 B2
Canterbury Rd,
Whitstable 5 C2
Canute Rd 10 A2
Capel Clo 15 C2
Carlton Hill 7 D1
Carlton Rd East 11 D1
Carlton Rd West 11 D1
Carlton Rd, Broadstairs 17 D2
Carlton Rd, Ramsgate 19 C3
Carlton Rise 11 D1
Caroline Cres 17 C2
Caroline Sq 13 D1
Carroways Pl 13 D1
Castle Av 17 D1
Castle Rd 5 C1
Castle Row 21 C3
Castle St 21 C3
Catherine Way 17 C2
Cattle Mkt 26 A2

Cavendish Rd 8 B1
Cavendish St 19 C2
Caxton Rd 13 C3
Cecil Park 8 B1
Cecil Sq 13 D1
Cecil St, Herne Bay 8 B1
Cecil St, Margate 13 D1
Cecilia Gro 17 C1
Cecilia Rd 19 C2
Cedar Clo, Cliftonville 14 A2
Cedar Clo,
Reading Street 17 C1
Cedar Rd 24 B2
Cedarview 20 A2
Cedric Rd 12 A2
Central Av 7 D1
Central Par 8 A1
Central Rd 19 C2
Chafy Cres 24 B2
Chalk Hill 18 A3
Chamberlain Dri 7 D1
Chanctonbury Chase 4 A3
Chandos Rd 17 D3
Chandos Sq 17 D3
Chapel Hill Clo 13 D3
Chapel La 24 A2
Chapel Pl, Margate 11 C2
Chapel Pl, Ramsgate 19 C2
Chapel Place La 19 C2
Chapel Rd 18 B2
Chapel Row 8 B3
Chapel St 8 B1
Charing Cres 11 D2
Charles Rd 19 C2
Charles St 8 B1
Charlesworth Dri 11 C2
Charlotte Pl 13 D2
Charlotte Sq 13 D2
Charlton Clo 18 B1
Charnwood Rd 8 B3
Chartwell Av 9 C1
Chatham Pl 19 C2
Chatham St 19 C2
Chaucer Av 6 A2
Chaucer Rd, Broadstairs 17 C3
Chaucer Rd, Canterbury 21 D3
Cheesmans Clo 25 B3
Cheriton Av 18 B2
Cherry Av 20 B2
Cherry Dri 20 B2
Cherry Garden Rd 20 B2
Cherry Gardens,
Herne Bay 8 A2
Cherry Gardens,
Northwood 16 B3
Cherry Gro 21 C2
Cherry Orchard 6 A3
Cherry Tree Gdns 16 A3
Chester Rd 12 A2
Chestfield Rd 6 A3
Chestnut Dri,
Broadstairs 16 B3
Chestnut Dri, Herne Bay 7 D2
Chestnut Dri, Sturry 24 B1
Cheviot Ct 17 D2
Chichester Rd 18 B1
Chilman Av 11 D2
Chilton La 18 B3
Chislett Way 15 C1
Chrysler Av 7 C1
Church Court Gro 16 B2
Church Hill,
Harbledown 20 A3
Church Hill, Ramsgate 19 C2
Church La, Canterbury 21 C3
Church La, Seasalter 4 A3
Church La, Sturry 24 A2
Church Rd, Broadstairs 17 D2
Church Rd, Margate 13 D2
Church Rd, Ramsgate 19 C2
Church Sq 13 D1
Church St, Canterbury 21 C3
Church St, Canterbury 20 B3
Church St, Margate 13 D2
Church St, Minster 26 A1
Church St, St Marys 26 A1
Church St, St Peters 17 C3
Church St, Sandwich 26 B2
Church St, Whitstable 5 D2
Church Way 6 B2
Church Wood Clo 20 A2
Churchfields 13 D2
Churchfields Pl 13 D2
Churchill Av 9 C1
Churchill Rd 23 D2
Churchwood Dri 6 B3

Citroen Clo 7 C1
City View 20 B3
Claire Ct 17 D2
Clapham Hill 5 C3
Clare Rd, Herne Bay 7 D2
Clare Rd, Whitstable 5 C1
Claremont Gdns 18 B2
Claremont Pl 22 B2
Claremont St 7 D2
Clarence Av 14 B2
Clarence Rd, Herne Bay 8 A1
Clarence Rd, Pegwell 18 B3
Clarence St 8 A1
Clarendon Gdns 19 C3
Clarendon Mews 17 C2
Clarendon Rd, Broadstairs 17 D3
Clarendon Rd, Cliftonville 14 A2
Clarendon St 7 D1
Cleaver La 19 C2
Clement Clo 9 C1
Clementine Clo 9 C1
Clements Rd 18 B1
Cliff Av 9 C1
Cliff Dri 7 D1
Cliff End 11 D1
Cliff Promenade 15 D3
Cliff Rd, Birchington 10 B1
Cliff Rd, Reading Street 17 D1
Cliff Rd, Whitstable 5 C1
Cliff St 19 C3
Cliff Ter 13 D1
Cliffe Av 12 B2
Clifford Rd 5 D2
Cliffsea Gro 7 D1
Clifftown Gdns 7 D1
Clifton Gdns, Canterbury 20 B2
Clifton Gdns, Margate 13 D1
Clifton Lawn 19 C3
Clifton Pl 14 A2
Clifton Rd, Margate 13 D1
Clifton Rd, St Lawrence 18 B2
Clifton Rd, Whitstable 5 C2
Clifton St 13 D1
Cliftonville Av, Cliftonville 14 A2
Cliftonville Av, Whitehall 18 B1
Cliftonville Mews 14 A1
Clive Rd 16 A2
Clovelly Rd 5 C3
Clover Rise 6 A3
Clowes Ct 20 A1
Clyde St 21 C2
Cobblers Bridge Rd 8 A2
Cobbs Pl 13 D1
Cobham Clo 23 D2
Cockering Rd 22 A3
Codringham Rd 4 B3
Codrington Rd 19 C3
Cogan Ter 22 B2
Colburn Rd 19 D1
Coleman Cres 18 B1
Colemans Stairs 11 C1
Colemans Stairs Rd 11 C1
Colette Clo 15 C2
Colewood Rd 6 B2
Collard Clo 8 B2
College Rd, Canterbury 21 D3
College Rd, Cliftonville 14 A3
College Rd, Margate 13 D2
College Rd, Ramsgate 19 C2
Collingwood Clo 17 C3
Collingwood Rd 5 C2
Collins Rd 7 D2
Colombo Sq 18 A1
Columbia Av 4 B3
Coniston Av 18 A2
Connaught Gdns 13 D2
Connaught Rd 13 D2
Conrad Av 21 D2
Constable Rd 10 B1
Consul Clo 7 C1
Convent Rd 17 C1
Convent Walk 18 B3
Conway Rd 10 B2
Conyngham Clo 18 B1
Conyngham Rd, Beltinge 9 C1
Conyngham Rd, Minster 25 B3
Coopers La 22 B2
Copinger Clo 21 C1
Copperhurst Walk 15 C2
Copt Clo 24 B1
Cornford Rd 11 C2
Cornhill 19 C3
Cornwall Av 19 D2
Cornwall Gdns, Canterbury 23 D2

Cornwall Gdns, Cliftonville 14 A2
Cornwall Rd 7 D2
Cornwallis Av 9 C2
Cornwallis Circle 5 C1
Cornwallis Gdns 17 D2
Coronation Clo 17 C1
Coronation Cres 12 B2
Coronation Rd 19 C3
Cossington Rd 23 C2
Cottage Rd 19 C3
Cottage Row 26 A1
Cotton Rd 22 B2
Coulter Rd 7 D2
Courtlands 8 B2
Courtlands Way 12 A2
Coventon La 26 A3
Coventry Gdns 9 D1
Cow La 22 B2
Cowdrey Pl 23 C2
Cowley Rise 14 B3
Cowper Clo 6 B2
Cowper Rd 13 D2
Coxes Av 16 A3
Coxes La 16 A3
Craddock Rd 21 D3
Cranbourne Clo 19 D2
Cranleigh Gdns 5 C2
Craven Clo 13 C3
Crawford Gdns 14 A2
Crawford Rd 17 C2
Crescent Rd, Birchington 11 C2
Crescent Rd, Kingsgate 15 D3
Crescent Rd, Margate 13 C2
Crescent Rd, Ramsgate 18 B2
Cresta Clo 7 C1
Crispe Rd 10 B3
Crofton Rd 12 A3
Crofts Pl 17 D2
Cromwell Rd, Canterbury 23 C2
Cromwell Rd, Whitstable 5 C1
Cross Rd 11 C1
Cross St, Canterbury 20 B3
Cross St, Herne Bay 8 A2
Crossley Av 7 C1
Crossways 21 C1
Crossways Av 16 A2
Crow Hill 17 D2
Crow Hill Rd 12 B3
Crown Gdns 20 B3
Crown Hill Rd 7 D1
Crundale Way 15 C2
Cudham Gdns 14 B2
Culpepper Clo 21 C1
Cumberland Av, Broadstairs 17 D2
Cumberland Av, Canterbury 23 D2
Cumberland Av, Cliftonville 14 A2
Cumberland Rd 19 C3
Cundishall Clo 4 B2
Cunningham Cres 10 B2
Curlinge Ct 18 B3
Cusham Rd 22 B2
Cuthbert Rd 12 A2
Cypress Clo 4 B3

Daimler Av 7 C2
Dalby Rd 14 A2
Dalby Sq 14 A2
Dallinger Rd 10 B1
Dalmaney Clo 17 D2
Dalmeny Av 14 B2
Dane Court Gdns 16 B2
Dane Court Rd 16 B1
Dane Cres 19 C2
Dane End Rd 11 D1
Dane Gdns 14 B3
Dane Hill 13 D1
Dane Hill Row 13 D1
Dane Mount 14 B3
Dane Park Rd, Margate 13 D1
Dane Park Rd, Ramsgate 19 C2
Dane Rd, Birchington 10 A2
Dane Rd, Margate 13 D1
Dane Rd, Ramsgate 19 C2
Dane Valley Rd 14 A3
Danesmead Ter 13 D1
Darenth Clo 8 B2
Darnley Clo 17 C3
Darrell Clo 7 D2
Darren Gdns 19 D1
Darwin Rd 10 B2
Daryngton Av 10 A2
David Av 14 B2
Davids Clo 17 D3

Davidson Rd 20 B2
Daytona Way 7 C1
Deal Rd 26 A3
Dean Croft 8 B3
Deane Clo 5 C2
Deansway Av 24 A1
Deborah Clo 5 D3
Delaware Clo 24 B2
Delf St 26 A1
Delfside 26 B2
Delmar Clo 6 B2
Denbigh Rd 18 A1
Dence Clo 8 B1
Dence Pk 8 B1
Denmark Rd 19 C2
Denne Clo 24 B1
Dent-de-Lion Rd, Garlinge 12 B3
Dent-de-Lion Rd, Westgate 12 A2
Dering Rd 8 A1
Derwent Av 18 A2
D'este Rd 19 D2
Detling Av 19 D1
Devon Gdns 10 B2
Devon Rd 23 D1
Devonshire Gdns 14 A2
Devonshire Ter 17 D2
Diamond Rd 5 C1
Dickens Av 21 D2
Dickens Rd 17 D2
Dickens Walk 17 D2
Doggeral Acre 5 D2
Dolphin Clo 15 C2
Dolphin St 8 A1
Domneva Rd, Minster 25 A2
Domneva Rd, Westgate 12 A2
Donegal Rd 21 D3
Donnahay Rd 16 B3
Dorcas Gdns 17 C1
Dorothy Dri 16 B3
Dorset Clo 4 B3
Dorset Gdns 10 B2
Dorset Rd 23 D2
Douglas Av 5 C2
Douglas Clo 17 C2
Douglas Rd 8 B2
Douro Rd 23 D1
Dove Clo 4 B3
Dover Rd 26 A3
Dover St 23 C1
Downs Av 5 C2
Downs Park 8 B1
Downs Rd, Pegwell 18 B3
Downs Rd, St Stephens 21 C1
Drapers Av 13 D2
Drapers Clo 16 A1
Drybeck Av 18 A2
Dryden Clo 23 D2
Duck La 21 C3
Dudley Av 11 D1
Duke St 13 D1
Dumpton Gap Rd 19 D1
Dumpton La 19 C1
Dumpton Park Dri 19 D1
Dumpton Park Rd 19 C2
Duncan Dri 10 B2
Duncan Rd, Ramsgate 19 C3
Duncan Rd, Whitstable 5 C2
Dundonald Rd, Broadstairs 17 D3
Dundonald Rd, Ramsgate 18 B2
Dunedin Rd 18 A1
Dunstan Av 11 D2
Durban Rd 14 A3
Durham Clo 23 C2
Durlock 25 B3
Durlock Av 18 B3
Durnford Clo 20 B2
Durovernum Ct 23 C2

Eagle Hill 19 C2
East Cliff Prom 17 D2
East Gate Clo 8 B2
East St, Canterbury 21 D2
East St, Herne Bay 8 B1
Eastchurch Rd 14 B2
Eastern Esplanade, Broadstairs 17 D2
Eastern Esplanade, Cliftonville 14 A1
Eastfield Rd 10 B2
Eaton Hill 13 D2
Eaton Rd 13 D2
Eddington La 8 A2
Eden Rd 4 A3
Edenfield 11 C2
Edgar Clo 6 B2
Edgar Rd, Canterbury 21 D3

Edgar Rd, Cliftonville 14 A1
Edgar Rd, Minster 25 A2
Edge End Rd 17 C3
Edinburgh Rd 12 B2
*Edinburgh Walk, Edinburgh Rd 12 B2
Edith Rd, Ramsgate 18 B3
Edith Rd, Westgate 12 A2
Edmanson Av 12 B2
Edmunds Clo 20 A2
Edward Dri 11 C2
Edward Rd 21 C3
Effingham St 19 C3
Egbert Rd, Birchington 10 A2
Egbert Rd, Minster 25 A2
Egbert Rd, Westgate 12 A2
Egerton Dri 15 C2
Elfrida Clo 14 B3
Elham Clo 14 B3
Elham Way 22 B2
Elham Way 19 D1
Elizabeth Rd 19 D3
Elizabeth Way 8 B2
Ellen Av 19 C1
Ellington Av 12 B3
Ellington Pl 18 B2
Ellington Rd 18 B2
Ellis Rd 5 D1
Ellis Way 9 C2
Ellison Clo 6 A3
Elm Gro, Canterbury 20 A2
Elm Gro, Westgate 12 A2
Elm Villas 20 B2
Elmdon Pl 17 D2
Elmley Way 13 D3
Elms Av 19 C3
Elmstone Gdns 15 C2
Elmstone Rd 19 C2
Elmwood Av 15 D3
Elmwood Clo, Reading Street 17 C1
Elmwood Clo, Swalecliffe 6 A2
Elmwood West 6 A2
Emmerson Gdns 6 A2
Empire Ter 13 D3
Enticott Clo 8 B3
Epping Clo 5 D2
Epple Bay Av 11 C1
Epple Bay Rd 10 B2
Epple Rd 11 C1
Ersham Rd 23 C2
Eskdale Av 18 B2
Essex Av 7 C1
Essex Gdns 10 B2
Essex Rd, Canterbury 23 D1
Essex Rd, Westgate 12 A2
Essex St 5 C2
Ethel Rd 17 C2
Ethelbert Cres 14 A1
Ethelbert Gdns 13 D1
Ethelbert Rd, Birchington 10 A1
Ethelbert Rd, Canterbury 23 C2
Ethelbert Rd, Margate 13 D1
Ethelbert Rd, Ramsgate 19 C3
Ethelbert Sq 12 A2
Ethelbert Ter, Margate 13 D1
Ethelbert Ter, Westgate 12 A2
Ethelred Rd 12 A2
Eynsford Clo 15 C2

Fair St 17 C3
Fairacre 17 C3
Fairacres Clo 9 C2
Fairfax Dri 9 D1
Fairfield Rd, Broadstairs 17 C3
Fairfield Rd, Minster 25 B2
Fairfield Rd, West Dumpton 19 C1
Fairlawn 6 B3
Fairlawn Rd 16 A3
Fairlight Av 18 B2
Fairoaks 9 C2
Fairview Gdns 24 B2
Farleigh Rd 21 C1
Farley Rd 13 D3
Farmhouse Clo 5 D2
Farrer Rd 11 C2
Faversham Rd 4 A3
Fern Clo 6 A3
Fern Ct 17 D2
Ferndale Ct 10 B2
Fernlea Av 8 A2
Field Av 21 D2
Field Way 24 B2
Fife Rd 7 D2
Fifth Av 14 A1
Fig Tree Rd 17 C1
Finsbury 19 C2

Fir Tree Clo, Ramsgate 18
Fir Tree Clo, Rough Common 20
Firbank Gdns 13
Firbanks 5
First Av, Cliftonville 14
First Av, Kingsgate 15
Fisher Rd 20
Fisher St 26
Fitzgerald Av 7
Fitzmary Av 12
Fitzroy Av, Cliftonville 14
Fitzroy Av, Kingsgate 15
Fitzroy Av, Whitehall 18
Fitzroy Rd 5
Fleetwood Av 8
Fletcher Rd 6
Flora Rd 19
Florence Av 13
Ford Clo 7
Fordoun Rd 17
Fordwich Gro 17
Fordwich Pl 26
Fordwich Rd 24
Foreland Av 14
Foreness Clo 15
Forge Clo 24
Forge Fields 8
Forge La 5
Forrester Clo 21
Fort Cres 13
Fort Hill 13
Fort Promenade 13
Fort Rd, Broadstairs 17
Fort Rd, Margate 13
Forty Acres Rd 20
Fosters Av 17
Fountain St 5
Foxborough La 25
Foxdown Clo 20
Foxgrove Rd 5
Francis Gdns 19
Francis Rd 17
Franklyn Rd 20
Freda Clo 19
Freemans Clo 4
Freemans Rd 25
Frencham Clo 21
Freshlands 9
Friars Clo 5
Friary Way 20
Friendly Clo 14
Friends Av 14
Fulham Av 12

Gainsborough Dri
Gainsborough Rd 10
Galliard St 26
Gallwey Av 10
Gann Rd 5
Garden Clo 20
Garfield Rd 13
Garrard Av 12
Gas St 21
Gateacre Rd 4
Genesta Av 4
George Hill Rd 15
George St 19
George V Av 12
Georges Av 4
Gilbert Rd 19
Gilchrist Av 7
Giles Gdns 13
Giles La 20
Gillon Mews 21
Gladstone Rd, Broadstairs 17
Gladstone Rd, Margate 13
Gladstone Rd, Whitstable 5
Glebe Gdns 12
Glebe Rd 12
Glebe Way 5
Glen Av 9
Glen Iris Av 20
Glen Iris Clo 20
Glenbervie Dri 9
Glenbrook Clo 9
Glencoe Rd 14
Glenside 5
Glenside Av 6
Gloucester Av, Broadstairs 17
Gloucester Av, Cliftonville 14
Gloucester Rd 5
Godwin Rd, Cliftonville 14
Godwin Rd, Thanington 22
Golden Acre Rd 11
Golden Clo 11
Golden Hill 5

Entry	Ref
oden Rd	21 C1
odwin Av	6 B2
odwin Rd	18 B3
don Rd, Canterbury	22 B2
don Rd, Cliftonville	14 A2
don Rd, Herne Bay	8 A2
don Rd, Ramsgate	19 C2
don Rd, Westwood	16 A2
don Rd, Whitstable	5 C2
don Sq	10 B2
e End Clo	10 B2
re Mews	21 D2
rrell Rd	5 C2
rse La	9 C2
sfield Rd	8 B1
sselin St	5 C2
udhurst Clo	21 C1
afton Rd	17 C1
afton Rise	7 D1
and Dil	7 D1
ange Rd, Beltinge	9 C1
ange Rd, Reading Street	17 C1
ange Rd, St Lawrence	18 B2
ange Way	19 D1
ant Clo	17 C2
anville Av, Broadstairs	17 D3
anville Av, Whitehall	18 B1
anville Dri	7 D2
anville Rd	17 D3
asmere Av	18 A2
asmere Rd	5 D2
avel Walk	21 C3
ays Way	22 A2
aystone Rd	5 D1
een Cloth Mews	21 D2
een Dell	21 C1
een La, Northdown	14 B3
een La, St Peters	17 C2
een Leas	6 B3
een Rd	10 B1
een St	5 C2
eenacres Clo	8 B2
eenfield	18 B1
eenhill Bridge Rd	7 D2
eenhill Clo	25 A2
eenhill Gdns, Herne Bay	7 D2
eenhill Gdns, Minster	25 A2
enham Bay Av	10 A2
enham Rd	10 B1
enville Gdns	10 B1
enville Way	17 C3
esham Av	12 B2
imshill Ct	20 A1
imshill Rd	5 C2
imthorpe Av	4 B3
osvenor Gdns	13 D2
osvenor Hill	13 D2
osvenor Pl	13 D1
osvenor Rd, Broadstairs	17 C3
osvenor Rd, Ramsgate	18 B2
otto Gdns	13 D1
otto Hill	13 D1
otto Rd	13 D1
ove Gdns	13 C2
ove Rd	19 C3
ove Ter	22 B2
ummock Rd	18 B2
undy's Hill	19 C3
uildcourt La	26 A1
uildford Av	12 A3
uildford Lawn	19 C3
uildford Rd	22 B2
uildhall St	21 C3
uy Clo	17 D1
wynn Rd	18 B1
ackington Clo	20 B1
ackington Pl	21 C2
ackington Ter	21 C2
adleigh Gdns	8 B1
adlow Dri	15 C2
aine Rd	18 A2
ales Dri	21 C2
alfmile Ride	13 D3
alford Clo	9 C3
allett Walk	21 D2
alstead Clo	21 C1
alstead Gdns	15 C2
am Shades La	6 D2
amilton Clo	18 B1
amilton Rd	5 C1
ampshire Rd	23 D2
ampton Gdns	7 C2
ampton Pier Av	7 D1
Hanover Clo	15 C2
Hanover Pl	21 C2
Hanover St	8 A1
Harbledown Gdns	14 B2
Harbledown Park	20 A3
Harbour Par	19 C3
Harb'ur St, Broadstairs	17 D2
Harbour St, Ramsgate	19 C3
Harbour St, Whitstable	5 C1
Harcourt Dri, Canterbury	20 B2
Harcourt Dri, Herne Bay	7 D1
Hardres Rd	19 C2
Hardres St	19 C2
Hardy Clo	20 B3
Harkness Dri	20 B2
Harmsworth Gdns	17 C2
Harnet St	26 A2
Harold Av	12 A2
Harold Rd, Birchington	10 A1
Harold Rd, Cliftonville	14 A2
Harrison Rd	19 C3
Harrow Dene	17 C2
Hartsdown Rd	13 C2
Harwich St	5 C2
Haslemere Rd	4 A3
Hastings Av	14 A2
Hatfield Rd	19 C2
Havelock St	21 C3
Haven Dri	9 D1
Hawe Clo	21 C1
Hawe Farm Way	8 B3
Hawe La	24 B1
Hawes Av	18 B2
Hawk Clo	4 B3
Hawkhurst Clo	11 C1
Hawkhurst Way	19 D1
Hawks La	21 C3
Hawks Rd	7 D2
Hawley Sq	13 D1
Hawley St	13 D1
Hawthorn Av	21 C2
Hawthorn Clo	19 C1
Hazelmere Dri	9 C1
Hazelwood Meadow	26 A2
Headcorn Dri	21 C1
Headcorn Gdns	15 C2
Heart-in-Hand Rd	9 D2
Heath Clo	24 A1
Heathersole Clo	21 C1
Heathwood Dri	19 C1
Heaton Rd	22 B2
Helding Clo	9 C2
Helena Av	13 D2
Helmdon Clo	16 B3
Helvellyn Av	14 A3
Hengist Av	17 C1
Hengist Rd, Birchington	10 A2
Hengist Rd, Westgate	11 B1
Herbert Rd	18 B3
Hereford Gdns	10 B2
Hereson Rd	19 D2
Hereward Av	10 A1
Herne Av	8 B2
Herne Bay Rd, Sturry	24 A1
Herne Bay Rd, Swalecliffe	6 A2
Herne Dri	7 D2
Herne St	8 B3
Herneville Gdns	8 B2
Herschell Rd	10 B1
Hertford Pl	19 C3
Hertford Rd	14 B3
Hertford St	19 C3
Hever Pl	21 C1
Hibernia St	19 C2
High Artillery Rd	21 C3
High St, Broadstairs	17 C3
High St, Canterbury	21 C3
High St, Fordwich	24 B3
High St, Garlinge	12 B2
High St, Herne Bay	8 A1
High St, Margate	13 D1
High St, Minster	25 B2
High St, Ramsgate	19 C2
High St, St Peters	17 C2
High St, Sandwich	26 B2
High St, Sturry	24 B2
High St, Whitstable	5 C1
High View Av	7 D1
Highbury Gdns	16 A3
Highfield Clo, Canterbury	20 A1
Highfield Clo, Northwood	16 A3
Highfield Gdns	13 C2
Highfield Rd	18 B1
Highfields Av	9 C1
Highfields Clo	9 C1
Highfields Vw	9 C1
Highgate Rd	6 A3
Hilary Clo	9 C1
Hildersham Clo	17 C2
Hill La	9 C2
Hill Top Rd	8 B1
Hill View Rd	5 C2
Hillborough Dri	9 D1
Hillborough Park	9 D1
Hillborough Rd	8 B1
Hillbrow Av, Hunters Forstal	8 B2
Hillbrow Av, Sturry	24 B1
Hillbrow Rd	19 C2
Hillcrest Gdns	18 B3
Hillcroft Rd	8 B2
Hillman Av	7 C1
Hillside Av	20 A2
Hillside Rd	5 D2
Hillview Rd	20 A2
Hinchcliff Way	14 B3
Hirst Dri	7 D2
Hoades Wood Rd	24 B1
Hobart Rd	18 A1
Hockeredge Gdns	12 A2
Hodgson Rd	4 A3
Hogarth Clo	9 D1
Hogs Corner	26 B2
Holbourn Clo	8 B3
Holbrook Dri	18 B1
*Holiday Sq, The Parade	13 D1
Holland Clo	15 D2
Hollicondane Rd	19 C2
Hollow La	22 B3
Hollowmede	22 B2
Holly Clo	16 B3
Holly Gdns	14 B2
Holly La	14 B2
Holly Rd	19 C2
Holm Oak Gdns	17 C3
Holmscroft Rd	9 C1
Homeleigh Rd	16 A3
Homestall Ct	20 A1
Homewood Rd	24 B2
Honeysuckle Rd	19 D2
Honeywood Clo	21 D2
Hoopers La	9 C2
Hopes La	16 B3
Hopeville Av	16 B2
Hornet Clo	16 B3
Horsa Rd	10 A2
Horsebridge Rd	5 C1
Hoser Gdns	11 C2
Hospital La	21 C3
Hovenden Clo	21 C1
Howard Rd	17 C3
Hubert Way	17 C1
Hudson Clo	24 B2
Hudson Rd	21 D2
Hugin Av	17 C1
Humber Av	7 C1
Hundreds Clo	11 D2
Hunter Forstal Rd	9 C2
Hunters Chase, Hunters Forstal	9 C3
Hunters Chase, Whitstable	5 C3
Hunting Gate	10 B2
Hurst Gro	18 B1
Ince Rd	24 B1
Ingle Clo	11 C2
Ingoldsby Rd, Birchington	10 A2
Ingoldsby Rd, Thanington	22 A2
Inverness Ter	17 D3
Invicta Rd, Cliftonville	14 A3
Invicta Rd, Whitstable	5 D2
Irchester St	5 C2
Irish Village	5 C2
Iron Bar La	21 C3
Irvine Dri	18 A2
Island Rd	24 B2
Island Wall	5 C1
Ivanhoe Rd, Herne Bay	8 B2
Ivanhoe Rd, Westgate	12 A2
Ivy La, Canterbury	21 C3
Ivy La, Ramsgate	19 C3
Ivy Ter	22 B2
Ivychurch Gdns	15 C2
Ivyhouse Rd	5 D2
John St	17 D3
Johns Green	26 A3
Joss Gap Rd	15 D2
Joy La	4 B3
Jubilee Rd	26 A2
Julie Clo	17 C1
Junction Rd	7 D2
Juniper Clo, Canterbury	23 C2
Juniper Clo, Whitstable	5 D2
Keith Av	18 A2
Kemp Rd	6 A2
Kemsing Gdns	21 D1
Kendal Clo	18 B3
Kendal Rise	17 C2
Kent Av	23 D1
Kent Gdns	10 B2
Kent Pl	19 D3
Kent Rd	14 A3
Kent St	5 C2
Kentmere Av	18 A2
Kenton Gdns	25 A2
Kevin Dri	18 B3
Keyworth Mews	21 D2
Kilndown Gdns, Canterbury	21 D1
Kilndown Gdns, Cliftonville	15 C2
Kimberley Gro	4 A3
Kimberley Rd	18 A1
King Edward Av	8 B1
King Edward Rd, Birchington	10 B3
King Edward Rd, Ramsgate	18 B3
King Edward St	5 C1
King Edwards Av	17 D3
King St, Canterbury	21 C3
King St, Margate	13 D1
King St, Ramsgate	19 C2
King St, Sandwich	26 B2
Kingfisher Clo, Garlinge	12 B3
Kingfisher Clo, Seasalter	4 B3
Kingfisher Ct	8 A2
Kings Av, Birchington	10 A2
Kings Av, Broadstairs	17 C3
Kings Av, Whitehall	18 B1
Kings Av, Whitstable	5 C2
Kings Rd, Birchington	11 C3
Kings Rd, Herne Bay	8 A1
Kings Rd, Ramsgate	19 C2
Kingsdown Park	5 D1
Kingsfield Rd	9 C2
Kingsgate Av	15 C2
Kingsgate Bay Rd	15 D2
Kingsley Rd	5 C2
Kingsmead Rd	21 C2
Kingston Av	12 B3
Kingston Clo, Beltinge	9 D1
Kingston Clo, Whitehall	18 B1
Kirbys La	21 C3
Kirkstone Av	18 A2
Kite Farm	6 B1
Knight Av	20 B3
Knightrider St	26 B2
Knights Av	17 D2
Knockholt Rd	15 C2
Knold Park	13 D3
Knotts La	21 C3
Knowler Way	9 C1
Knowlton Walk	21 D2
*La Belle Alliance Sq, Plains of Waterloo	19 C2
Laburnham La	24 B1
Laburnum Av	26 A2
Lady Woottons Grn	21 C3
Ladysmith Gro	4 A3
Ladywood Rd	24 B1
Lagos Av	18 B1
Laking Av	17 D1
Laleham Gdns	14 A2
Laleham Rd	14 A3
Lamberhurst Way	15 C2
Lambeth Rd	21 D1
Lambs Walk	4 B3
Laming Rd	11 C2
Lancaster Clo	18 B1
Lancaster Gdns, Beltinge	9 C1
Lancaster Gdns, Birchington	10 B2
Lancaster Rd	22 B2
Lanchester Clo	9 C1
Landon Rd	8 A1
Lane End	8 B2
Lanfranc Gdns	20 A3
Langdale Av	18 B2
Langham Clo	12 B2
Langley Gdns	15 C2
Langton La	23 C3
Lansdown Rd	23 C2
Lanthorne Rd	17 C1
Larch Clo	16 B3
Latimer Clo	7 C2
Laundry Rd	25 B2
Laureate Clo	14 A2
Lauriston Clo	18 B3
Lauriston Mount	17 C2
Lausanne La	13 D2
Lausanne Ter	13 D2
Lawley Clo	18 B1
Lawn Rd	17 D2
Lawrence Gdns	9 C2
Laxing Av	17 D1
Laxton Way	6 A3
Laylam Clo	16 B2
Leas Grn	16 B3
Leatt Clo	17 C3
Lerryn Gdns	15 C3
Lesley Av	23 C2
Leslie Av	12 B3
Leslie Rd	11 C1
Lewis Cres	14 A1
Leybourn Rd	17 D3
Leybourne Dri	12 B2
Leycroft Clo	21 C2
Lichfield Av	23 D2
Lillian Rd	19 D2
Lime Kiln Rd	22 B2
Lincoln Av	23 C2
Lincoln Clo	6 B2
Lincoln Gdns	10 B2
Linden Av, Broadstairs	17 D2
Linden Av, Herne Bay	8 A2
Linden Av, Whitstable	5 D2
Linden Clo	12 A3
Linden Gro	21 C3
Linden Rd	12 A3
Lindenthorpe Rd	17 C2
Linington Rd	11 C2
Link Dri	17 C1
Link La	21 C3
Links Clo	8 B2
Linksfield Rd	12 A3
Linley Rd	17 C1
Linnet Av	4 B3
Linton Gdns	21 C1
Lismore Rd, Beltinge	9 C1
Lismore Rd, Swalecliffe	6 A3
Lister Rd	14 A3
Liverpool Lawn	19 C3
Livingstone Rd	17 C2
Lloyd Rd	17 C2
Lombard St	13 D1
London Rd, Canterbury	20 B3
London Rd, Ramsgate	18 B3
Long Meadow Way	21 C1
Long Reach Clo	4 B3
Long Rock	6 B2
Longacre	6 B3
Longmarket	21 C2
Longmead	7 D2
Longport	21 D3
Longtye Dri	6 A3
Lonsdale Av	14 A2
Loop Ct Mews	26 A1
Loop St	26 A2
Lorina Rd	18 B1
Lorne Rd	18 B3
Love La, Canterbury	21 C3
Love La, Margate	13 D1
Love La, Sandwich	26 B2
Love Street Clo	7 D2
Lovell Rd	20 A2
Lower Bridge St	21 C
Lower Chantry La	23 C1
Lower Herne Rd	8 B3
Lower Northwood Av	14 A2
Lucerne Dri	4 A3
Luckhurst Gdns	15 C2
Luton Av	17 C3
Luton Clo	17 C3
Lyell Rd	10 B2
Lymington Rd	12 A3
Lyndhurst Av	18 B1
Lyndhurst Rd, Broadstairs	17 D2
Lyndhurst Rd, Ramsgate	19 D2
Lytham Av	8 A3

McCarthy Av 24 B1
Macdonald Par 4 A3
Madeira Rd 14 A2
Madeira Walk 19 C3
Magdala Rd 17 C2
Magdalen Ct 17 D2
Magnolia Av 14 B2
Mallory Clo 18 B1
Malthouse Rd 21 C2
Malvern Pk 9 C1
Mandeville Rd 20 B2
Maney Gdns 17 D2
Manor Clo 22 A3
Manor Dri, Beltinge 9 D1
Manor Dri, Birchington 10 B2
Manor Rd, Beltinge 9 D1
Manor Rd, Tankerton 5 D1
*Mansion St, The Parade 13 D1
Manston Rd, Birchington 11 C3
Manston Rd, Margate 13 C3
Manston Rd, St Lawrence 18 A2
Manwood Av 21 C1
Manwood Rd 26 B2
Maple Clo 20 A2
Marden Av 18 B2
Margate Hill 11 C3
Margate Rd, Herne Bay 9 C2
Margate Rd, Westwood 16 A2
Marilyn Cres 11 C2
Marina Rd 19 C2
Marine Cres 6 A2
Marine Dri, Kingsgate 15 D2
Marine Dri, Margate 13 D1
Marine Esplanade 19 D3
Marine Gdns 13 D1
Marine Parade 5 D1
Marine Ter, Margate 13 C2
Marine Ter, Seasalter 4 B1
Mark Av 18 B3
Mark Green Walk 21 D2
Market Sq, Herne Bay 8 A1
Market Sq, Margate 13 D1
Market Way 21 C2
Marlborough Clo 17 C3
Marlborough Rd, Margate 13 D2
Marlborough Rd, Ramsgate 19 C3
Marley Ct 20 B1
Marlow Clo 6 A2
Marlowe Av 21 C3
Marlowe Rd 14 B3
Marrose Av 16 A3
Marsh Farm Rd 25 B3
Marshall Cres 17 C3
Marshwood Clo 21 D2
Martindale Clo 23 C2
Martindown Rd 4 B3
Martins Clo 18 B1
Martyrs Field Rd 22 B2
Masons Rise 17 C2
Matthews Rd 7 D2
Maugham Ct 5 C2
Maxine Gdns 17 C2
Maydowns Rd 6 B2
Mayfield Rd 8 B2
Mayforth Gdns 18 B3
Maynard Av 12 B2
Maynard Rd 22 B2
Mays Rd 18 B3
Maystreet 9 D1
Mayton Rd 24 A1
Mead Way 20 B3
Meadow Clo 9 C1
Meadow Dri 6 B3
Meadow Rd, Canterbury 20 A3
Meadow Rd, Sturry 24 B2
Meadow Rd, Westgate 12 B2
Meadow Walk 4 B3
Medina Av 4 B3
Meeting St 19 C2
Melbourne Av 18 B1
Mellanby Clo 11 C2
Melsetter Clo 11 C2
Mentmore Rd 16 B3
Mercury La 21 C3
Merchants Way 20 A3
Mertons La 22 B3
Metcalfe Mews 21 D2
Meteor Av 4 B2
Michael Av 19 D2
Michelle Gdns 12 B2
Mickett Dri 17 C1
Mickleburgh Av 8 B2
Mickleburgh Hill 8 B1

Middle Wall 5 C1
Military Rd, Canterbury 21 C3
Military Rd, Ramsgate 19 C3
Mill Clo 26 A1
Mill Cotts 19 C3
Mill La, Birchington 10 B2
Mill La, Canterbury 21 C3
Mill La, Harbledown 20 A3
Mill La, Hunters Forstal 8 B2
Mill l a, Margate 13 D2
Mill rd 24 A3
Mill Row 10 B2
Mill Stream Clo 5 C2
Miller Av 20 B3
Millfield 17 C2
Millfield Manor 5 C2
Millfield Rd 16 A3
Millmead Av 14 B3
Millmead Gdns 14 B3
Millmead Rd 14 A3
Millstrood Rd 5 D2
Millview Rd 8 B3
Millwall Pl 26 B2
Milner La 24 A2
Milner Rd 4 A3
Milton Av 13 D2
Milton Clo 23 D2
Milton Rd 23 C2
Minnis Rd 10 A1
Minster Clo 19 D1
Minster Dri 8 A1
Minster Rd, Pegwell 18 B3
Minster Rd, Westgate 12 A3
Moat Sole 26 A2
Molehill Rd 6 B3
Molineux Rd 25 A3
Monastery St 21 C3
Monkton Gdns 15 C2
Monkton Rd 25 A2
Montague Rd 19 C2
Montana St 8 A1
Montefiore Av 19 D1
Montfort Clo 21 C1
Montpelier Av 5 C3
Moorfields 21 C1
Moray Av 10 B1
Mordaunt Av 12 A2
Morris Av 7 C1
Mortimer St 8 A1
Mount Rd 23 D2
Mount View Rd 8 B3
Mountfield Way 11 D2
Muir Rd 19 D2
Mulberry Clo 19 D2
Mutrix Gdns 12 B2
Mutrix Rd 12 B2
Mymms Clo 6 A3

Nacholt Clo 5 D1
Nackington La 23 C2
Nackington Rd 23 C3
Napier Rd 17 C2
Napleton Rd 18 B3
Nash Court Gdns 13 D3
Nash Court Rd 13 D3
Nash Gdns 17 D2
Nash La 13 D3
Nash Rd, Margate 13 D3
Nash Rd, Westwood 16 A2
Nasmyth Rd 11 C1
Neals Place Rd 20 A2
Neame Rd 11 C2
Nelson Cres 19 C3
Nelson Ct 10 B1
Nelson Pl 17 D2
Nelson Rd 5 C2
Nethercourt Farm Rd 18 B2
Nethercourt Gdns 18 B3
Nethercourt Hill 18 B3
Nethersole Clo 21 C1
New Cross St 13 D1
New Dover Rd 23 C2
New House La 22 B3
New Ruttington Pl 21 C3
New St, Canterbury 20 B3
New St, Herne Bay 8 B1
New St, Margate 13 D1
New St, Martyrs Field 22 B2
New St, Sandwich 26 A2
New Town St 21 C2
*Newbys Pl, Market St 13 D1
Newcastle Hill 19 C2
Newgate Lower Prom 14 A1
Newgate Promenade 14 A1
Newington Rd 10 B1
Newlands La 16 B3
Newlands Rd 19 C1
Newton Rd 6 A2
Nicholls Av 19 D1

Nightingale Av 4 B3
Nixon Av 18 B1
Noble Gdns 12 B2
Norfolk Rd, Canterbury 22 B2
Norfolk Rd, Cliftonville 14 A2
Norfolk St 5 C2
Norman Rd, Canterbury 23 C2
Norman Rd, Ramsgate 18 B3
Norman Rd, Westgate 12 A2
Norman Rd, Whitstable 5 C2
Norrie Rd 10 B2
North Av 19 C3
North Foreland Av 15 D3
North Foreland Hill 15 D3
North Foreland Rd 15 D3
North Holmes Rd 21 D3
North La 21 C3
North St 8 B1
Northcliffe Gdns 17 C1
Northdown Av 14 A2
Northdown Hill 16 B1
Northdown Park Rd 14 A2
Northdown Rd, Margate 13 D1
Northdown Rd, Reading Street 17 C1
Northdown Way 14 B3
Northgate 21 C3
Northumberland Av 14 B2
Northwood Rd, Northwood 16 B3
Northwood Rd, Whitstable 5 C1
Norton Av 8 B3
Norton Dri 25 B2
Norview Rd 4 B3
Notley St 21 C2
Nottingham Rd 10 B3
Nunnery Fields 23 C2
Nunnery Rd 23 C2
Nursery Clo, Church Street 5 D2
Nursery Clo, St Lawrence 18 B2

Oak Tree Gro 12 B2
Oakdale Rd 8 B2
Oakdene Rd 18 B1
Oakfield Ct 18 B2
Oaklands Av 16 B2
Oakridge 15 C2
Oaks Dri 7 D2
Oaks Park 20 A1
Oakwood Dri 5 D1
Oakwood Rd 24 B1
Oast Ct 13 D2
Oatenhill 23 C2
Oatenhill Pl 23 C2
Ocean Clo 11 C1
Ocean View, Beltinge 9 D1
Ocean View, Dumpton 19 D1
Offley Clo 14 B2
Olave Rd 14 A3
Old Boundary Rd 12 A2
Old Bridge Rd 5 C2
Old Crossing Rd 12 B2
Old Dover Rd 23 C2
Old Farm Clo 5 C3
Old Farm Rd 10 A2
Old Fold 6 A3
Old Green Rd, Cliftonville 14 B2
Old Green Rd, Reading Street 17 C1
Old Harbour 21 C2
Old Park Av 21 D2
Old Ruttington La 21 C3
Omer Av 14 B2
Orange St 21 C3
Orchard Clo, Minster 25 B2
Orchard Clo, Northwood 16 A3
Orchard Clo, St Stephens 21 C1
Orchard Clo, Tankerton 5 D1
Orchard Gdns 12 B2
Orchard Rd, Herne Bay 8 A2
Orchard Rd, Westgate 12 B2
Orchard Row 8 B3
Orchard St 20 B3
Osborne Gdns 9 C1
Osborne Rd 17 C3
Osborne Ter 13 D2
Osbourn Av 12 A2
Oscar Rd 17 D3
Osprey Clo 4 B3
Otham Clo 21 C1
Owls Hatch Rd 7 D3
Oxenden Park Dri 8 A2
Oxenden Sq 8 A1

Oxenden St 8 A1
Oxford Rd 22 B2
Oxford St, Margate 13 D2
Oxford St, Whitstable 5 C2
Oxney Clo 10 B2

Paddock Rd 10 B2
Paddock View 5 C3
Palace Clo 6 A2
Palace St 21 C3
Palm Bay Av 14 B1
Palm Bay Gdns 14 B1
Palmer Cres 14 B3
Palmers Cross Hill 20 A2
Palmerston Av 17 D3
Paradise 19 C2
Paradise Row 26 A1
Paragon 19 C3
Paragon St 19 C3
Parham Rd 21 D2
Park Av, Birchington 10 B2
Park Av, Dumpton 19 C1
Park Av, Tankerton 5 D1
Park Chase 19 C1
Park Crescent Rd 13 D2
Park Gate 19 D1
Park La, Birchington 11 C2
Park La, Margate 13 D2
Park Pl, Herne 8 B3
Park Pl, Margate 13 D2
Park Pl, Birchington 11 C2
Park Rd, Broadstairs 17 D2
Park Rd, Herne Bay 8 A1
Park Rd, Margate 13 D2
Park Rd, Ramsgate 19 C2
Park View 24 A2
Parkland Ct 17 C2
Parkwood Clo 19 D1
Parsonage Rd 8 B2
Patricia Way 16 B3
Payton Clo 13 D3
Payton Mews 21 C2
Pear Tree Clo 16 B3
Pearsons Way 16 B1
Peartree Rd 9 C2
Pegwell Av 18 B3
Pegwell Clo 18 A3
Pegwell Rd 18 A3
Pembroke Av 12 B2
Penshurst Clo 21 C1
Penshurst Gdns 15 D2
Penshurst Rd 19 D2
Percy Av 15 C2
Percy Rd, Broadstairs 17 C2
Percy Rd, Cliftonville 14 A1
Percy Rd, Ramsgate 19 C2
Perkins Av 13 D2
Petchell Mews 21 D2
Pettman Clo 8 B2
Petts Cres 25 A3
Phillips Rd 11 C2
Picton Rd 18 B3
Pier App 17 D2
Pier Av, Herne Bay 8 A1
Pier Av, Tankerton 5 D1
Pier Chine 8 A1
Piermont Rd 5 C3
Pierremont Av 17 D2
Pigeon La 8 B2
Pilgrims Rd 23 D2
Pilgrims Way 23 D2
Pilkem Clo 23 D1
Pin Hill 21 C3
Pine Tree Clo 5 D1
Pine Walk 9 D1
Pinetree Av 20 B2
Pinewood Clo 18 B1
Plains of Waterloo 19 C2
Plantation Rd 6 B2
Pleydell Cres 24 B1
Plough Ct 9 C2
Plough La 6 B1
Pluckley Gdns 15 C2
Poets Corner 13 D2
Polo Way 6 A3
Pondicherry All 26 B2
Poorhole La 16 A2
Popes La 24 A1
Poplar Dri 7 D2
Poplar Rd, Ramsgate 19 C2
Poplar Rd, St Peters 17 C2
Potter St 26 B2
Poulders Gdns 26 A2
Poulders Rd 26 A2
Pound La 21 C3
Powell Cotton Dri 11 C2
Prescott Way 8 B3
Prestedge Av 19 C1
Preston Par 4 A3

Pretoria Rd 2
Prices Av, Cliftonville 1
Prices Av, Ramsgate 1
Priest Av 2
Priest Walk
Priestfields
Prince Andrew Rd
Prince Charles Rd 1
Prince Edwards Promenade 1
Princes Av 1
Princes Clo 1
Princes Gdns 1
Princes Rd 1
Princes St 1
Princes Walk 1
Princes Way 2
Princess Anne Rd 1
Princess Clo
Princess Cres 1
Princess Margaret Av, Newington 1
Princess Margaret Av, Northdown 1
Prioress Rd 2
Priory Clo 1
Priory La
Priory Rd 1
Priory of St Jacob 2
Prospect Clo 1
Prospect Gdns 2
Prospect Hill
Prospect Pl, Broadstairs 17
Prospect Pl, Canterbury 23
Prospect Rd, Birchington 1
Prospect Rd, Broadstairs
Prospect Rd, Minster 2
Prospect Ter 1
Providence Row 2
Puckle La 2
Pullman Clo 1
Pysons Rd 1

Quantock Gdns 1
Quay La 2
Queen Bertha Rd 1
Queen Berthas Av 1
Queen Elizabeth Av 1
Queen St, Herne Bay 8
Queen St, Ramsgate 1
Queens Av, Beltinge 9
Queens Av, Birchington
Queens Av, Broadstairs 17
Queens Av, Canterbury 20
Queens Av, Margate 13
Queens Av, Whitehall 18
Queens Gate Rd 18
Queens Gdns, Broadstairs 17
Queens Gdns, Herne Bay 8
Queens Gdns, Margate 14
Queens Promenade 14
Queens Rd, Broadstairs 17
Queens Rd, Ramsgate 19
Queens Rd, Tankerton 5
Queens Rd, Westgate 12
Queensbridge Dri 7
Querns Pl 21
Querns Rd 21
Quetta Rd 18
Quex Rd 12
Quex View Rd 10

Radley Clo 17
Radnor Clo 8
Raglan Pl 17
Railway Av 5
Railway Ter 13
Ramsey Clo 20
Ramsgate Rd, Broadstairs 17
Ramsgate Rd, Malgate 13
Ramsgate Rd, Sandwich 20
Rancorn Rd 13
Randolph Clo 23
Ranelagh Gro 17
Ravensbourne Rd 8
Ravenscourt Rd 20
Rawdon Rd 18
Rayham Rd 5
Raymond Av 23
Reading St 17
Reading Street Rd 17
Rectory Gdns 6
Rectory Rd 17

ulver Av 10 B2
ulver Clo 9 D1
ulver Dri 9 D1
ulver Rd 9 C2
ulvers Rd 12 A3
d Lion La, Sea St 5 C1
cot La 24 B1
dhill Rd 12 A3
dwood Clo 20 B2
d Av 21 D2
ency Clo 5 D2
gent St 5 C1
nston Mews 21 D2
nault Clo 7 C1
oton Clo 17 C2
servoir Rd 5 C1
ynolds Clo 8 B1
eims Way 20 B3
odaus Clo 22 B2
odaus Town 21 C3
odes Gdns 17 C1
hborough Rd, Sandwich 26 A1
hborough Rd, Westgate 12 A3
hmond Av 14 A2
hmond Dri 9 D1
hmond Gdns 20 B2
hmond Rd, Ramsgate 19 C3
hmond Rd, Swalecliffe 6 A3
hmond St 8 A1
geway 6 A3
geway Cliff 7 D1
lley Clo 8 B3
ey Av 7 C1
gold Av 18 A2
gwood Clo 20 B2
don Clo 24 B2
erdale Rd 21 D2
erside Rd 18 B1
erview 24 A1
berts Rd 4 A3
chester Av 23 C2
ckingham Pl 8 B3
ckstone Way 18 A1
dney St 19 C3
man Rd 18 B1
milly Gdns 18 B1
mney Clo 10 B2
per Clo 20 B2
per Rd 21 C2
pers La 8 A3
se Gdns, Beltinge 9 C1
se Gdns, Birchington 10 B2
se Gdns, Minster 25 A2
se Hill 19 C3
se La 21 C3
seacre Clo 20 B2
sebery Av, Beltinge 9 D1
sebery Av, Ramsgate 19 D1
sedale Rd 14 A2
selands Gdns 20 B2
selea Av 8 B2
semary Av 17 C3
semary Gdns, Broadstairs 17 C3
semary Gdns, Whitstable 5 D2
semary La 21 C3
setower Ct 17 C1
ss Gdns 20 A2
ssetti Rd 10 B2
ssland Rd 18 A2
ugh Common Rd 20 A1
wan Clo 24 B2
wena Rd 12 A2
wland Cres 9 D1
wland Dri 7 D2
xburgh Rd 12 A2
yal Cres 13 C2
yal Esplanade, Ramsgate 18 B3
yal Esplanade, Westgate 12 B2
yal Par 19 C3
yal Rd 19 C3
gby Clo 17 C2
mfields Rd 16 B2
shmead Clo 20 B2
ssell Dri 6 B2
tland Av 14 A2
tland Clo 23 D2
tland Gdns, Birchington 10 B2
tland Gdns, Cliftonville 14 B2
dal Av 18 A2
Ryde St 20 B3
Ryders Av 11 D1
Rye Walk 9 C2
Sacketts Hill 16 A1
Saddlers Mews 6 B3
Saddleton Rd 5 C2
St Alphege Clo 4 B3
St Alphege La 21 C3
St Andrews Clo, Herne Bay 8 B2
St Andrews Clo, Margate 13 D3
St Andrews Clo, Whitstable 5 C3
St Andrews Lees 26 B2
St Andrews Rd 19 D2
St Annes Dri 8 A2
St Annes Gdns 13 D3
St Annes Rd 5 D1
St Anthonys Way 14 B2
St Augustines Av 13 D3
St Augustines Cres 6 B2
St Augustines Pk 18 B3
St Augustines Rd, Canterbury 23 C2
St Augustines Rd, Ramsgate 19 C3
St Barts Rd 26 A2
St Benedicts Lawn 19 C3
St Benets Rd 12 A3
St Christopher Clo 14 B3
St Clements Rd 12 A2
St Crispins Rd 12 A3
St Davids Clo, Birchington 11 C2
St Davids Clo, Whitstable 5 C2
St Davids Rd 19 D1
St Dunstans 20 B2
St Dunstans Clo 20 B2
St Dunstans Rd 14 A2
St Dunstans St 20 B2
St Dunstans Ter 20 B3
St Edmunds Rd 21 C3
St Francis Clo 14 B3
St Georges Av 7 D1
St Georges La 21 C3
St Georges Lees 26 B2
St Georges Pl 23 C1
St Georges Rd, Broadstairs 17 D3
St Georges Rd, Ramsgate 19 D2
St Georges Rd, Sandwich 26 B2
St Georges St 21 C3
St Georges Ter, Canterbury 21 C3
St Georges Ter, Herne Bay 8 A1
St Gregorys Rd 21 D3
St Jacobs Pl 22 B2
St James Av, St Peters 17 C2
St James Av, Whitehall 18 B1
St James Gdns 5 C2
St James Park Rd 12 B2
St James Ter 11 C2
St Jeans Rd 12 A3
St Johns Av East 18 A1
St Johns Av West 18 A1
St Johns La 21 C3
St Johns Pl 21 C3
St Johns Rd, Margate 13 D2
St Johns Rd, Swalecliffe 6 B2
St Johns St 13 D2
St Julien Av 23 D1
St Lawrence Av 18 B3
St Lawrence Clo 23 C2
St Lawrence Forstal 23 C2
St Lawrence High St 18 B2
St Lawrence Rd 23 C2
St Louis Gro 7 D1
St Lukes Av 19 C2
St Lukes Clo 12 A3
St Lukes Rd 19 C2
St Magnus Clo 10 B1
St Magnus Ct 10 C1
St Margarets Clo 4 A3
St Margarets Rd 12 A3
St Margarets St 21 C3
St Marks Clo 5 C2
St Martins Av 21 D3
St Martins Clo 21 D3
St Martins Hill 21 D3
St Martins Pl 21 D3
St Martins Rd 21 D3
St Martins Ter 21 D3
St Martins View 8 B3
St Marys Av 14 B3
St Marys Gro 4 A3
St Marys Rd, Broadstairs 17 D2
St Marys Rd, Minster 25 B2
St Marys St 21 C3
St Michaels Av 14 B3
St Michaels Clo 20 A2
St Michaels Pl 20 B2
St Michaels Rd 20 B2
St Mildreds Av, Birchington 10 B2
St Mildreds Av, Broadstairs 17 C3
St Mildreds Av, Ramsgate 18 B3
St Mildreds Gdns 12 A2
St Mildreds Pl 22 B2
St Mildreds Rd, Cliftonville 14 A2
St Mildreds Rd, Minster 25 B3
St Mildreds Rd, Ramsgate 18 B3
St Mildreds Rd, Westgate 12 A2
St Nicholas Clo 24 B1
St Nicholas Rd 22 A2
St Patricks Rd 19 D2
St Pauls Rd 14 A2
St Peters Ct 17 C2
St Peters Footpath 13 D2
St Peters Gro 21 C3
St Peters La 21 C3
St Peters Pl 21 C3
St Peters Rd, Broadstairs 17 C2
St Peters Rd, Margate 13 D2
St Peters St, Canterbury 21 C3
St Peters St, Sandwich 26 B2
St Peters St, Whitstable 5 C1
St Radigunds Pl 21 C3
St Radigunds St 21 C3
St Stephens Clo 21 C2
St Stephens Ct 21 C2
St Stephens Footpath 21 C2
St Stephens Grn 21 C2
St Stephens Hill 20 B1
St Stephens Rd 21 C2
St Swithins Rd 6 A2
St Thomas Hill 20 B2
Salisbury Av, Broadstairs 17 C3
Salisbury Av, Ramsgate 19 C2
Salisbury Rd, Beltinge 9 C1
Salisbury Rd, St Stephens 20 B2
Salisbury Rd, Whitstable 5 C2
Salmestone Rd 13 D2
Salts Dri 17 C2
Saltwood Gdns 14 B2
Sancroft Av 20 B2
Sanctuary Clo 19 D1
Sandhurst Clo 21 C1
Sandhurst Rd 15 C2
Sandles Rd 10 B2
Sandown Dri 8 A1
Sandown Rd 26 B2
Sandpiper Rd 4 B3
Sandwich Rd 18 A3
Sandwood Rd, Ramsgate 19 D1
Sandwood Rd, Sandwich 26 A2
Sangro Pl 23 D1
Sarah Gdns 14 B3
Savernake Dri 8 B3
Saxon Rd, Pegwell 18 B3
Saxon Rd, Westgate 12 A2
School La, Herne 8 B3
School La, Ramsgate 19 C2
School Rd 26 A1
Sea App 17 D3
Sea Rd 11 C1
Sea St, Hampton 7 D1
Sea St, Whitstable 5 C1
Sea View Av 10 B1
Sea View Rd 9 C1
Sea View Rd, Birchington 10 B1
Sea View Rd, Broadstairs 17 D2
Sea View Sq 8 A1
Sea View Ter 13 C2
Sea Wall 5 C1
Seacroft Rd 19 D1
Seafield Rd, Broadstairs 17 C3
Seafield Rd, Ramsgate 18 B2
Seafield Rd, Swalecliffe 6 A2
Seamark Rd 10 B3
Seapoint Rd 17 D3
Seasalter Beach 4 B2
Seasalter La 4 A3
Seaville Dri 9 D1
Second Av, Cliftonville 14 A1
Second Av, Kingsgate 15 C2
Seeshill Clo 5 C2
Selborne Rd 14 A3
Selsea Av 7 D1
Selwyn Dri 17 C2
Semaphore Rd 10 B1
Senlac Clo 18 B3
Setterfield Rd 13 D2
Sevastapol Pl 23 D1
Seven Stones Dri 19 D1
Sewell Clo 11 C2
Seymour Av, Westgate 12 B2
Seymour Av, Whitstable 5 C2
Seymour Pl 22 B2
Shaftesbury Rd, Canterbury 21 C2
Shaftesbury Rd, Whitstable 5 C1
Shaftesbury St 19 D2
Shah Pl 19 C2
Shakespeare Passage 13 C2
Shakespeare Rd 10 B1
Shalloak Rd 24 A2
Shallows Rd 16 B1
Shamrock Av 4 B3
Shapland Clo 9 C2
Share & Coulter Rd 6 B2
Shearwater Av 4 B3
Shell Gdns 7 D2
Shelley Av 21 D2
Shepherds Walk 6 A3
Sheppey Clo 10 B2
Sheppey View 4 B3
Sherwood Clo, Hunters Forstal 9 C3
Sherwood Clo, Seasalter 4 B3
Sherwood Dri 4 B3
Sherwood Gdns 19 C1
Sherwood Rd 10 B3
Shipman Av 20 B3
Shirley Av 19 C1
Short St 26 B2
Shottendane Rd 12 A3
Shutler Rd 17 D2
*Sidney Pl, Market St 13 D1
Silver Av 11 C3
Silverdale Rd 18 A3
Simmonds Rd 22 B2
Simon Av 14 B2
Singer Av 7 C1
Sion Hill 19 C3
Slades Clo 6 A3
Sleigh Rd 24 B2
Sloe La 16 A2
Smugglers Way 11 C1
Sobraon Way 23 D1
Somerset Clo 4 B3
Somerset Rd 23 D1
Somme Ct 23 D1
Somner Clo 20 B2
Sondes Clo 8 B2
South Canterbury Rd 23 C2
South Cliff Parade 19 D1
South Eastern Rd 18 B3
South Lodge Clo 5 C1
South Rd 8 B1
South St, Canterbury 21 D2
South St, Whitstable 5 D2
South View Rd 5 C3
Southsea Dri 8 A2
Southwold Pl 12 A3
Southwood Gdns 18 B2
Southwood Rd, Ramsgate 18 B2
Southwood Rd, Swalecliffe 6 A2
Sowell St 17 C2
Spa Esplanade 7 D1
Speke Rd 17 C2
Speldhurst Gdns 15 C2
Spencer Rd, Birchington 10 B1
Spencer Rd, Herne Bay 8 A2
Spencer Sq 19 C3
Spire Av 5 D2
Spratling La 18 A1
Spratling St 18 A1
Spring La, Canterbury 23 D2
Spring La, Sturry 24 B3
Spring Walk 5 C3
Springfield Clo 19 C1
Springfield Rd 15 C2
Squire Av 20 B3
Staffordshire St 19 C2
Staines Hill 24 B1
Staines Pl 17 D2
Stancomb Av 18 B3
Standard Av 7 C1
Staner Ct 18 A2
Stanley Gdns 8 B2
Stanley Pl, Broadstairs 17 D2
Stanley Pl, Ramsgate 19 C2
Stanley Rd, Broadstairs 17 C2
Stanley Rd, Cliftonville 14 A2
Stanley Rd, Herne Bay 8 B2
Stanley Rd, Ramsgate 19 C2
Stanley Rd, Whitstable 5 C2
Stanmore Ct 23 C2
Staplehurst Av 19 D1
Staplehurst Gdns 15 C2
Star La 16 A2
Starle Clo 21 D2
Station App, Birchington 10 B2
Station App, Minster 25 B3
Station App, Ramsgate 19 C2
Station Chine 8 A2
Station Par 10 B2
Station Rd East 22 B2
Station Rd, Birchington 10 B2
Station Rd, Canterbury 21 C3
Station Rd, Herne Bay 8 A2
Station Rd, Margate 13 C2
Station Rd, Minster 25 B3
Station Rd, Westgate 12 A2
Station Rd, Whitstable 5 C1
Stephen Clo 17 D3
Stephens Clo, Garlinge 12 B3
Stephens Clo, Ramsgate 18 B2
Stephenson Rd 21 C2
Sterling Clo 17 C2
Stirling Way 18 A1
Stockbury Gdns 15 C2
Stockwell Ter 20 A2
Stonar Clo, Dumpton 19 C1
Stonar Clo, Sandwich 26 B1
Stonar Gdns 26 B1
Stone Barn Av 11 C2
Stone Cross Lees 26 A3
Stone Gdns 17 D2
Stone Rd 17 D2
Stour St 21 C3
Strand St 26 A1
Strangers Clo 22 A2
Strangers La 22 A2
Strangford Pl 8 B3
Strangford Rd 5 D1
Stream Walk 5 C1
Streete Court Rd 12 A2
Streete Ct 12 A2
Streetfields 8 B3
Strode Park Rd 8 B3
Stuart Ct 23 C2
Stuppington La 22 B3
Sturry Hill 24 A2
Sturry Rd 21 D2
Sudbury Pl 12 A3
Suffolk Av 12 A3
Suffolk Rd 23 D1
Suffolk St 5 C2
Summer Hill 20 B3
Summerfield Av 5 D2
Summerfield Rd 15 C2
Sun St 21 C
Sunbeam Av 7 C1
Sundew Gro 19 C2
Sundridge Clo 21 C1
Sunningdale Walk 7 D3
Sunnyhill Rd 7 D1
Sunnyside Gdns 26 A2
Sunray Av 4 B3
Sunset Clo 4 B3
Surrey Gdns 10 B2
Surrey Rd, Canterbury 23 D2
Surrey Rd, Cliftonville 14 A2
Sussex Av, Canterbury 23 D2
Sussex Av, Margate 13 D2
Sussex Clo 7 D1
Sussex Gdns, Birchington 10 B2
Sussex Gdns, Hampton 7 D1
Sussex Gdns, Westgate 12 A2
Sussex St 19 C2
Swakeley Walk 6 A2
Swale Clo 8 B2
Swalecliffe Av 7 C1
Swalecliffe Court Dri 6 A2
Swalecliffe Rd 6 A2
Swallow Av 4 B3
Swanfield Rd 5 C2
Sweechbridge Rd 9 D2
Sweechgate 24 A1
Sweyn Rd 14 A2
Swinburne Av 17 C3

Swinford Gdns 14 B3
Sycamore Clo, Beltinge 9 C1
Sycamore Clo, Northwood 16 B3
Sydenham St 5 C1
Sydney Cooper Clo 20 A2
Sydney Rd, Ramsgate 19 D2
Sydney Rd, Whitstable 5 C2

Taddy Gdns 14 B3
Talavera Rd 23 D1
Talbot Av 7 C2
Talbot Rd 14 A2
Tankerton Mews 5 D1
Tankerton Rd 5 C1
Tassell Walk 6 B2
Tavistock Rd 19 C1
Taylor Rd 25 A2
Teddington Clo 21 D2
Telford St 8 A1
Telham Av 18 B2
Temple Rd 20 B3
Tennyson Av 21 D2
Tenterden Dri 21 C1
Terminus Dri 9 C1
Terrys La 5 C1
Teynham Clo 15 C2
Teynham Rd 5 D1
Thanet Clo 17 D2
Thanet Place Gdns 17 D1
Thanet Rd, Broadstairs 17 D2
Thanet Rd, Margate 13 D1
Thanet Rd, Ramsgate 19 D2
Thanet Rd, Westgate 12 A2
Thanet Way 4 B3
Thanington Rd 22 A2
The Banks 17 C2
The Borough 21 C3
The Bridge App 5 D1
The Broadway, Broadstairs 17 C2
The Broadway, Hampton 7 D1
The Butchery 26 B1
The Causeway, Canterbury 21 C2
The Causeway, Sandwich 26 A1
The Centre 18 A1
The Chain 26 B2
The Circus 8 A2
The Cloisters 19 C3
The Close 21 C1
The Courts 12 B2
The Crescent, Canterbury 21 C1
The Crescent, Sandwich 26 A3
The Dene 23 C3
The Downings 8 B2
The Drive, Canterbury 23 C2
The Drive, Chestfield 6 B3
The Drove, Chestfield 6 B3
The Drove, Sturry 24 B3
The Fairway 7 D2
The Foreland 23 C3
The Friars 21 C3
The Gap 23 C3
The Grange 4 A3
The Grove, Greenhill 7 D2
The Grove, Westgate 12 A2
The Halt 5 D2
The Hawthorns 16 B3
The Heath 6 A3
The Heights 4 B3
The Horshams 9 C1
The Hoystings Clo 23 C2
The Leas 6 A3
The Lees 9 C1
The Maples 16 B3
The Meadows 9 C2
The Ness 23 C3
The Nook 16 B3
The Oaks 17 C1
The Paddock 21 D3
The Paddocks, Beltinge 9 D1
The Paddocks, Reading Street 17 C1

The Parade, Birchington 10 A2
The Parade, Broadstairs 17 D2
The Parade, Margate 13 D1
The Pathway 17 D2
The Pines 16 B3
The Promenade 17 D3
The Quay 26 B2
The Rendezvous 13 D1
The Retreat, Birchington 11 C1
The Retreat, Whitehall 18 B2
The Ridgeway, Broadstairs 17 C3
The Ridgeway, Cliftonville 14 A2
The Ridings 6 B3
The School Clo 12 A2
The Silvers 16 B3
The Square 11 C2
The Terrace 21 C1
The Vale 17 C3
The Warren 5 C3
The Warren Drive 11 D2
Third Av 14 A1
Thirlmere Av 18 A2
Thorn Gdns 19 C1
Thornden Clo 7 C2
Thornden Ct 20 A1
Thornden Wood Rd 7 C2
Thorne Rd 25 A2
Throwley Dri 7 D1
Thundersland Rd 8 B2
Thurlow Av 9 C1
Thurston Pk 5 C2
Tina Gdns 17 D1
Tippledore La 17 C2
Tivoli Park Av 13 C2
Tivoli Rd 13 D2
Tollgate Clo 5 C2
Tomlin Dri 14 B3
Tomsons Passage 19 C2
Tonford La 22 A2
Tothill St 25 B2
Tower Gdns 8 A1
Tower Hill 5 C1
Tower Par 5 C1
Tower Rd 5 C1
Tower Way 21 C3
Townley St 19 C3
Trinity Hill 13 D1
Trinity Pl 19 D2
Trinity Sq, Margate 13 D1
Trinity Sq, Reading Street 17 C1
Truro Rd 19 D2
Tudor Clo 11 C2
Tudor Ct 20 B1
Tudor Rd 22 B2
Tunis Ct 23 D1
Tunis Row 17 D2
Tunstall Rd 21 C1
Turnden Gdns 15 C2
Turner St 19 C2
Twyne Clo 24 B2
Tydemans Av 6 A3
Tyler Clo 21 C1
Tyler Way 6 B2
Tyndal Park 8 B1
Tyson Av 12 B2

Ulcombe Gdns 21 C1
Ulster Rd 13 D2
Underdown La 8 B2
Underdown Rd 8 B1
Union Cres 13 D1
Union Pl 21 C3
Union Row 13 D1
Union Sq 17 D2
Union St, Canterbury 21 C3
Union St, Ramsgate 19 C2
Unity Pl 19 D2
Uplands 21 C1
Upper Approach Rd 17 D3
Upper Bridge St 21 C3
Upper Chantry La 23 C2
Upper Dane Rd 14 A2

Upper Dumpton Park Rd 19 C2
Upper Freedown 8 B2
Upper Gro 13 D1
Upper Strand St 26 B2
Upton Rd 17 C2
Ursuline Dri 11 D2

Vale Rd, Broadstairs 17 C3
Vale Rd, Ramsgate 19 C3
Vale Rd, Whitstable 5 C2
Vale Sq 19 C3
Valkyrie Av 4 B2
Valley Rd 22 B2
Vauxhall Av, Canterbury 21 D1
Vauxhall Av, Hampton 7 C2
Vauxhall Cres 21 D1
Vauxhall Rd 21 D1
Vere Rd 17 D2
Vereth Rd 19 C3
Vernon Pl 23 C2
Verwood Clo 20 B2
Vestey Ct 12 A2
Vicarage Cres 13 D2
Vicarage La 26 A1
Vicarage Pl 13 D2
Vicarage St 16 B2
Victor Av 14 B2
Victoria Av, Cliftonville 14 A2
Victoria Av, Reading Street 16 B1
Victoria Av, Westgate 12 A3
Victoria Dri 7 D1
Victoria Par, Broadstairs 17 D3
Victoria Par, Ramsgate 19 D2
Victoria Park 8 B1
Victoria Rd, Canterbury 22 B2
Victoria Rd, Margate 13 D2
Victoria Rd, Ramsgate 19 D2
Victoria Rd, St Peters 17 C2
Victoria Row 21 C3
Victoria St 5 C1
Viking Clo 10 A2
Villiers Rd 23 D1
Vincent Clo 16 B3
Vine Clo 19 C1
Violet Av 16 B3
Virginia Rd 5 D2
Vulcan Clo 5 C2

Wacher Clo 21 C2
Waldron Rd 19 D1
Wallace Way 17 C3
Wallwood Rd 19 D2
Walmer Rd 5 C2
Walmsley Rd 17 C2
Walnut Tree Clo 11 C2
Walpole Rd 13 D1
Waltham Clo 15 C2
Wantsume Lees 26 A1
Wardour Clo 17 D2
Warre Av 18 B3
Warren Dri 17 C2
Warten Rd 19 D1
Warwick Dri 18 B3
Warwick Rd, Canterbury 23 D1
Warwick Rd, Cliftonville 14 A2
Warwick Rd, Whitstable 5 C1
Watchester Av 18 B3
Watchester La 25 A3
Waterloo Pl 19 C2
Waterloo Rd 5 C1
Waterside Dri 12 A2
Watling St 21 C3
Wauchope Rd 4 A3
Wave Crest 4 B2
Waverley Rd 13 C2
Wayne Clo 17 C2
Wayville Rd 17 C1
Weigall Pl 18 B2
Well Clo 24 B2
Wellesley Clo, Broadstairs 17 C3
Wellesley Clo, Westgate 12 A3
Wellesley Clo, Cliftonville 14 A2

Wellesley Rd, Westgate 12 A3
Wellington Clo 12 A2
Wellington Cres 19 D3
Wellington Rd 12 A3
Wellis Gdns 13 C2
Wells Av 23 D2
Welsdene Rd 12 B3
Wemyss Way 23 D1
Wentworth Av 12 B2
Wentworth Dri 18 A1
Wentworth Gdns 8 A2
West Beach 4 B2
West Cliff 5 C2
West Cliff Av 17 D3
West Cliff Dri 7 D1
West Cliff Prom, Broadstairs 17 D3
West Cliff Prom, Ramsgate 19 C3
West Cliff Rd, Broadstairs 17 D3
West Cliff Rd, Ramsgate 19 C3
West Cliff Ter 18 B3
West Dumpton La 19 C1
West Hill Rd 7 D1
West Park Av 14 B2
West Pl 20 B2
West View Clo 7 D2
Westbrook Av 12 B2
Westbrook Cotts 13 C2
Westbrook Gdns 13 C2
Westbrook La 7 C2
Westbrook Prom 13 C2
Westbrook Rd 13 C2
Westbury Rd 12 A2
*Westcliff Arcade, Rose Hill 19 C3
Westcliff Gdns 12 B2
Westcliff Rd 13 C2
Westerham Clo 21 C1
Westerly Dri 19 D1
Western Av 8 A1
Western Esplanade, Broadstairs 17 D3
Western Esplanade, Hampton 7 D1
Western Rd 14 B3
Western Undercliff 18 B3
Westfield Rd, Birchington 10 B2
Westfield Rd, Westgate 12 B2
Westgate Bay Av 12 A2
Westgate Clo 20 B3
Westgate Court Av 20 A3
Westgate Gro 21 C3
Westgate Hill Rd 21 C3
Westgate Ter 5 C1
Westlands Rd 7 D2
Westleigh Rd 11 D1
Westmeads Rd 5 C1
Westminster Rd 21 D1
Westonville Av 12 B2
Westover Gdns 17 C1
Westover Rd 17 C1
Westwood Rd 16 A2
Weyburn Dri 18 B1
Wharfdale Rd 14 A2
Whatmer Clo 24 B2
Wheatley Rd, Whitehall 18 B1
Wheatley Rd, Whitstable 5 B1
Whinfell Av 18 A2
White Horse La 21 C3
Whitefriars Meadow 26 A2
Whitefriars Way 26 A2
Whitehall Bridge Rd 20 B3
Whitehall Clo 20 B3
Whitehall Gdns 20 B3
Whitehall Rd, Canterbury 20 B3
Whitehall Rd, Whitehall 18 B2
Whiteness Grn 15 C2
Whiteness Rd 15 C3
Whitfield Av 18 B3
Whitstable Rd, Canterbury 20 A1
Whitstable Rd, Hampton 7 C2
Wichling Clo 21 C1

Wickham Av 19
Wife of Bath Hill 20
Wilbrough Rd 11
Wilderness Hill 14
Wilfred Rd 18
Wilkes Rd 17
Wilkie Rd 10
William Av 14
William St 8
Willow Av 16
Willow Clo, Canterbury 21
Willow Clo, Cliftonville 14
Willow Grn 16
Willow Tree Clo 9
Willow Way 6
Willows Ct 20
Willsons Rd 19
Wilton Rd 18
Wimborne Pl 18
Wincheap 22
Winchester Gdns 23
Windermere Av 18
Windmill Av 16
Windmill Clo 21
Windmill Rd, Canterbury 23
Windmill Rd, Herne 8
Windmill Rd, Seasalter 5
Windsor Av 14
Windsor Clo 16
Windsor Rd 22
Wings Clo 17
Winifred Av 18
Winstanley Cres 19
Winston Clo 11
Winston Gdns 9
Winterstoke Cres 19
Winterstoke Undercliff 19
Winterstoke Way 19
Wolseley Av 7
Wood Vale 6
Woodford Av 19
Woodford Ct 11
Woodland Av 11
Woodland Way, Canterbury 20
Woodland Way, Kingsgate 15
Woodlawn St 5
Woodman Av 6
Woodnesborough Rd 26
Woodrow Cres 8
Woodside 24
Woodville Clo 22
Woodville Rd 18
Woolletts Clo 7
Wraik Hill 4
Wrentham Av 7
Wrotham Av 17
Wrotham Cres 17
Wrotham Rd 17
Wye Gdns 15
Wyndham Av 14
Wynn Rd 5

Yalding Clo 21
Yardhurst Gdns 15
Yarrow Clo 17
Yew Tree Clo 16
Yew Tree Gdns 18
Yoakley Sq 13
York Av 5
York Clo 8
York Rd, Canterbury 22
York Rd, Herne Bay 8
York St, Broadstairs 17
York St, Ramsgate 19
York Ter, Ramsgate 19
York Ter, Westgate 11
Ypres Ct 23

Zealand Rd 22
Zion Pl 13